HOW TO CHOOSE AND USE AN
AIR COMPRESSOR

MKTG 779

HOW TO CHOOSE AND USE AN
AIR COMPRESSOR

A complete handbook describing the air compressor and its many tools used in and around the home, containing more than three hundred and fifty photographic illustrations and line drawings.

Prepared for

CAMPBELL HAUSFELD

by

Robert Scharff & Associates, Ltd.
RD 1
New Ringgold, Pennsylvania 17960

FOREWORD

A recent survey revealed that one of the major problems the do-it-yourselfer has is to know "how to choose and use an air compressor." To rectify this situation, Campbell-Hausfeld, in their role as leaders of the air compressor industry, prepared this book.

As the title states, the purpose of this book is to help you select the right compressor for your needs, and once selected, how to get the most from it. The book describes how an air compressor makes *many* jobs around the house a breeze. It details what ways a compressor can make your car and motorized equipment run better and last longer. The air compressor is invaluable to every hobbyist, whether an artist, gardener, ceramist, model maker, woodworker, or outdoorsperson. By referring to this book as a guide, you will see how an air compressor system will pay for itself many times over regardless of your interests by saving time, effort and money.

We would like to thank Rockwell International; Chicago Pneumatic Tool Company; Star Dental Manufacturing Company, Inc; Badger Air-Brush Company; Briar Industries, Inc.; Spotnails, Inc.; and the 3M Company for the use of certain tools and accessories that appear in this book. The editors also wish to thank Thomas Draper, Arthur Roell, Duane Hall, and the many other employees of Campbell-Hausfeld for their untiring efforts in obtaining and checking the material in this book.

It is our sincere hope that, after reading this book, you will be able to choose and use an air compressor to your best advantage.

CAMPBELL ⊕ HAUSFELD

Division of the Scott & Fetzer Company
Harrison, Ohio 45030

Copyright © 1979 Campbell-Hausfeld,
Division of the Scott & Fetzer Company.

Brief quotations may be used in critical articles and reviews. For any other reproduction of this book, including electronic, mechanical, photocopying, recording or other means, written permission must be obtained from the publisher.

Library of Congress Catalog Card Number: 79-89642
Manufactured in the United States of America

CONTENTS

Foreword iv
Chapter 1 - What Is A Compressor? 1
Chapter 2 - Air Tool Accessories 13
Chapter 3 - Operation Of A Compressor 29
Chapter 4 - Basics Of Spray Painting 61
Chapter 5 - Outdoor Uses For An Air Compressor 81
Chapter 6 - Interior Uses For An Air Compressor 109
Chapter 7 - The Air Compressor And Your Car 121
Chapter 8 - The Air Compressor And Your Hobbies 137
Index 152

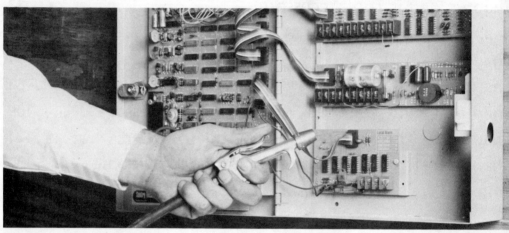

Fig. 1-1: Some of the many uses of an air compressor.

Chapter 4
BASICS OF SPRAY PAINTING

The spraying of paint and other finishing materials, as mentioned earlier, is one of the major uses of an air compressor system. There are many good reasons for this. An air-powered sprayer reduces the time and effort required to produce the lasting, professional finish that your work around the home deserves. Once the basic techniques of using a finish materials sprayer are learned, you will achieve a better finishing job than many skilled workers could do by hand. Actually, the sprayer is not a difficult tool to master. You will be amazed at the speed with which you can spray finish even large projects. The difference between a sprayed finish and a hand-brushed one is immediately obvious and is very rewarding to accomplish. With practice, skill in spray coating with excellent uniformity in thickness and appearance can be achieved. The option to use an air compressor-driven spray system permits the coating of common surfaces—high ceilings, wooden patio decks, and large and small objects, for example—which are more difficult to reach or completely inaccessible with the paint brush.

In this chapter, we are concerning ourselves with the basic techniques of paint spraying. In other chapters, specific procedures such as exterior spray painting, workshop spraying, repainting of your car or truck, and other projects will be fully detailed. But, before doing any spraying, review these safety suggestions carefully:

1. Carefully read the owner's manual before operating the spray gun, and observe the manufacturer's instructions.
2. Never smoke while spraying.
3. Provide ample ventilation while spraying indoors.
4. Never spray near open flames or pilot lights in stoves or heaters.
5. Wear a respirator (Fig. 4-1), particularly when spraying indoors.

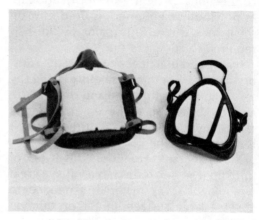

Fig. 4-1: Two styles of respirators.

SPRAY SYSTEMS

As described in Chapter 2, a finishing spray system consists of three major parts: a compressor, a hose, and a gun. The compressor delivers air through the hose to the gun. Here, the air picks up the fluid, atomizes it, and sprays it onto the work. A filter and regulator usually control the amount of air pressure. The gun can be used with either a canister system or with a pressure material tank.

Siphon Canister Hookup. The canister can be either pressure or siphon feed. A gun is actually set up as a siphon feed when drawing from a material tank. As shown in Fig. 4-2, connect the filter as close to the gun as possible. Be sure that all the hose

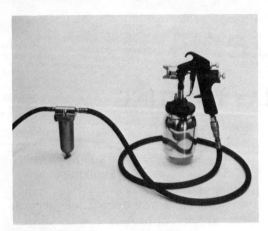

Fig. 4-2: Siphon canister hookup.

SPRAY PREPARATION

The prerequisite of a professional spray finish for any job is making sure that the surface is dry and free of all foreign materials such as dirt, grease, rust, oil, and loose paint. It is true that some paints and finishes are more tolerant than others to the presence of foreign matter; however, nearly all coatings will benefit in durability by applying them to a truly clean, smooth surface. In many cases, this rigid requirement will mean the difference between success and failure of the coating job.

The best possible method of removing dirt, rust, and scale from metal is sandblasting (Fig. 4-4). Where sandblasting is not feasible, wire brushing, scraping, or chipping will usually remove any metal contaminants. The best way to clean and smooth wooden surfaces is by sanding. Fortunately for the compressor owner, pneumatic tools are available that will perform all of these tasks. Details on how to use sandblast guns, metal cleaning tools, and wood sanders are covered in Chapters 5, 7, and 8.

connections are tight. This type of hookup, as previously mentioned, is used only on small jobs where little paint or finish is required to complete the job.

It is a good practice to turn ON the compressor for a few minutes before doing any spraying. Attach the hose and then let the air pass through it for several minutes to clear it and blow out any moisture. After it is clear, attach the gun.

Material Tank Hookup. This type of hookup (Fig. 4-3) is desirable when a great quantity of spray material is needed to cover a large surface, or for production spraying. The regulator (or filter regulator) on the compressor controls the pressure going to the spray gun. The regulator on the pressure material tank controls the amount of pressure with which the spray hits the material.

Fig. 4-4: Sandblasting is one of the best methods of removing rust.

The second prerequisite of a good spray finish is the proper preparation of the spray material whether it is paint, lacquer, varnish, shellac, etc. All the elements of the finish material, including pigments, vehicle, or thinner or solvent, must be thoroughly combined. When a fresh paint can is opened, for example, you will notice

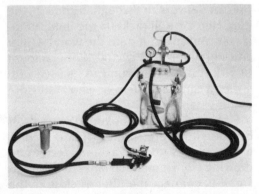

Fig. 4-3: Material tank hookup.

that some vehicle-thinner fluid is at the top. Mixing the pigment when the container is full will result in spilling a great deal of the vehicle and thinner, and will make proper mixing next to impossible.

To mix a full container of paint, pour off a portion of the vehicle into a clean container; then the remaining pigment and vehicle can be mixed with a paddle without spilling. After the pigment has been mixed thoroughly with this portion of the vehicle, mix the spray material further by pouring it from one container to another. Do this several times until the consistency of the material is uniform.

Next, add thinners as indicated on the container label, or use the chart on page 72. Most materials in common use will spray readily if thinned according to the manufacturer's directions for brushing. However, if the material still appears viscous, add a little more thinner, determining the amount by testing the mixture in a gun. As a general rule, spitting is a sign of thick paint, running means the paint is too thin. A guide is to thin the spray material to the consistency of SAE-20 motor oil.

The viscosity of the finishing material can be measured accurately by using a viscosimeter (Fig. 4-5). The device has a small hole in the bottom to measure how long the material remains in the cup. Just dip it into the finishing material, remove a full cup, then count the number of seconds it takes for the cup to empty. The following table indicates the time it takes for various finishing materials at 72 degrees F. to pass through a typical viscosimeter:

TYPE OF MATERIAL	TIME IN SECONDS
Auto enamel	25
Epoxy enamel	35
Glossy enamels	30
Interior latex	35
Lacquer	20
Latex floor enamel	30
Latex house paint	35
Oil house paint	35
Oil wall paint	30
Varnish	25

To be sure that the spray material is clean and free from lumps, strain it through a fine-mesh wire screen (60 to 80 mesh). An excellent substitute for a fine-mesh screen is an old nylon stocking (Fig. 4-6).

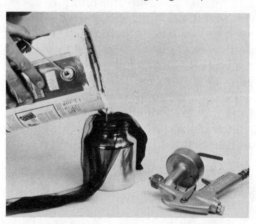

Fig. 4-6: Straining paint through a nylon stocking.

USE OF THE SPRAY GUN

Although handling a spray gun is simple, it is wise to spray a few practice panels before undertaking the actual work. Practice spraying can be done on old cartons or on sheets of newspaper tacked to a panel (Fig. 4-7). Ten or fifteen minutes of practice work will give you the feel of the gun and enough know-how to tackle an actual job.

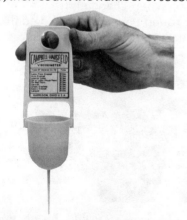

Fig. 4-5: A viscosimeter in use.

Fig. 4-7: The best way to learn the effects of gun movement and gun adjustments is to experiment on a test surface.

If possible, do a few simple jobs before attempting anything complicated or anything that necessitates skillful gun handling and flawless results.

Gun Canister Spraying. To practice spraying techniques, fill the canister about one-quarter full with the finishing material. Set the air pressure as desired; 45 to 50 psi is a good starting point. Start the compressor. Have the sprayer near enough to the work to allow a full movement of the gun at 6 to 9 inches from the newspaper or cardboard—about equal to the distance from thumb to middle finger of your outstretched hand (Fig. 4-8). Trigger the gun and check the spray. If it is too fine, reduce the air pressure or open the fluid control. If the spray is too coarse, close the fluid control. Also notice how the spray material begins to sag and finally run as you hold the gun motionless. This illustrates a very important point: never hold the gun motionless while it is spraying, or the fluid will continue to pile up on that one spot.

The spray pattern (Fig. 4-9) depends on the air cap type and how it is adjusted. The fan-shaped pattern is used for spraying flat wide surfaces, while the round pattern is preferred for small, irregular surfaces and close work. To change the fan direction for vertical or horizontal strokes, loosen the

Fig. 4-9: The standard fan and round spray patterns.

air cap and turn the nozzle. With some gun models (Fig. 4-10), the spray shape can be changed from a round to a fan-shaped pattern by turning the pattern control knob. When changing from a fan-shaped to a round pattern, the fluid control knob must be readjusted. Regardless of the

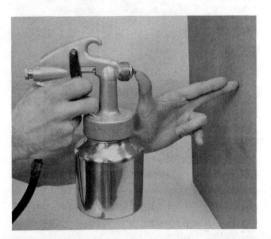

Fig. 4-8: Proper gun distance.

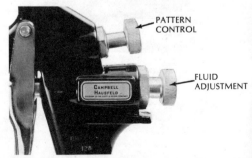

Fig. 4-10: Some spray guns have both the pattern and the fluid control at the rear.

method of controlling the pattern, when the length of the fan shape and the amount of paint is increased, the spray material particles become coarser until it becomes difficult to spray the heavy, hard-to-control pattern. The ideal is an easily handled pattern that still delivers enough spray material to efficiently cover the surface.

Work speed is important to the appearance of the finish. On becoming familiar with how the gun handles and feels, back down the fluid control until you get a very small pattern and then try moving the gun across the paper. The speed with which the gun moves controls the uniformity of the application. Moving too fast is not good, because the paint does not cover. Moving too slowly is even worse, since you will deposit too much paint and it will sag and run, ruining the job.

Material Tank Spraying. The only difference in spraying technique between canister and material tank sprayers is the method of loading the tank. When doing this, be sure to thoroughly mix and strain the material (Fig. 4-11) to remove skins or undissolved particles which might otherwise impede the flow of the material through the hose and gun. Plastic liners are

Fig. 4-11: A fine mesh screen can be used to strain the finish material as it is poured into a tank. Paint strainers are available that can be used in place of the screen.

available that fit inside the tank and hold the paint. Once the paint job is completed, the liner is removed and can either be cleaned or thrown away. These tank liners simply clean up. For smaller jobs, a gallon bucket of paint may be placed inside the paint tank and the paint used directly from the bucket. When this is done, the lid must be clamped down in the usual manner. No change in connections is necessary.

After the tank (Fig. 4-12) is filled, attach the lid by engaging the T bolts in the slotted lugs on the lid and tightening each wing nut a little at a time, making sure the lid is pulled down evenly and the seal between the tank and the lid is airtight. Shut off paint tank regulator by turning handle counterclockwise. Adjust the regulator on the compressor to obtain the desired air pressure on the spray gun. Now adjust the regulator on the paint tank to obtain the desired pressure on the material. The higher above the paint tank you are spraying the more pressure you will need on the material. Normal operating pressure on the paint tank is 25 to 30 psi (Fig. 4-13). Should you wish to reduce pressure, simply rotate the adjustment stem counterclockwise until the desired pressure setting is obtained. There is no need to trigger the gun in order to bleed off excess paint pressure. Caution: Do not use over 60 psi air pressure in your tank.

Most manufacturers recommend 15 feet lengths of air and material hose be used between the material tank and the spray gun. Any desired length of hose may be used between the material tank and the compressor. The handle of the material tank may be attached to the rung of a ladder or scaffolding, making some projects, such as spray painting sides of houses or barns, much easier. Using medium viscosity paint, the spray gun can be a maximum of approximately 15 feet above the pressure material tank with the compressor remaining at ground level.

When test spraying on cardboard or newspaper, adjust the gun in the same

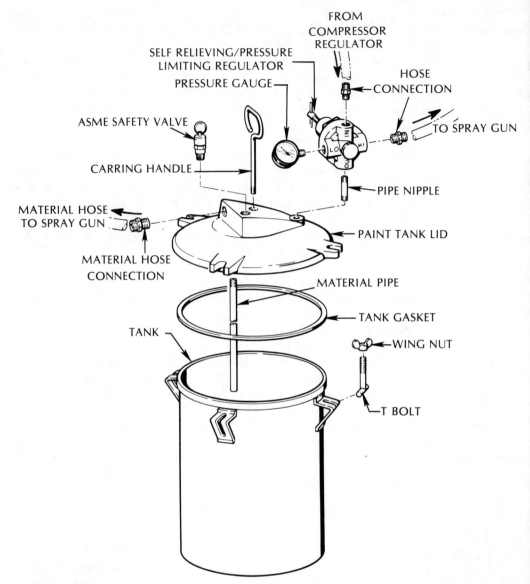

Fig. 4-12: Major parts of material tank.

manner that was described in using a canister. It will not be too long before you find yourself working with a good pattern and automatically feeding the gun across the work to deliver the right amount of finish material uniformly.

Stroking and Triggering The Gun

The spray gun stroke is made by moving the gun parallel to the work (Fig. 4-14) and at a right angle to the surface by flexing the wrist at each end of every stroke. Never allow a stiffly held wrist to "arc" the stroke. This is a fault common to beginners. Arcing the gun (Fig. 4-15) results in uneven application and excessive overspray at each end of the stroke. When the gun is arched 45 degrees away from the surface, for example, approximately 65 percent of the material is lost. Also when the gun is tilted, more material will be deposited at one edge of

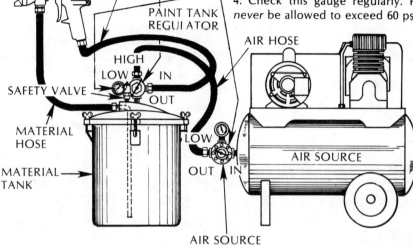

START-UP AND OPERATION
1. Shut off this regulator. (Turn handle counter-clockwise.)
2. Adjust this regulator to pressure required for your spray gun and paint tank regulator.
3. Adjust this regulator to 25-30 psi. (Turn handle clockwise slowly until gauge reads 25-30 psi.)
4. Check this gauge regularly. Pressure should never be allowed to exceed 60 psi maximum.

Fig. 4-13: A material tank in hookup and operation. The tank illustrated is equipped with a safety valve set at the factory. Never tamper with this valve.

the pattern (Fig. 4-16).

The distance from the gun to the work should be between 6 and 9 inches. The closer the gun is held to the work, the more paint is deposited on the work surface; the more paint applied, the faster the gun must be moved to prevent sags (Fig. 4-17A). When the gun is too far from the work, the stroke must be slowed down to a speed that is beyond average patience; the distant position is also undesirable because the paint may dry before it reaches the work surface, causing dusting as shown in Fig. 4-17B.

Triggering each stroke becomes a habit as you develop skill. Correct triggering technique prevents a buildup of material along the edges of the work. Start each stroke with the trigger OFF. Depress the trigger at the edge of the work and continue the stroke to the opposite edge; then release the trigger and follow through with

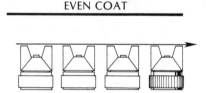

Fig. 4-14: Keep the gun parallel to the work surface.

Fig. 4-15: Arcing the stroke causes poor distribution of the finish, with too much at the center of each stroke.

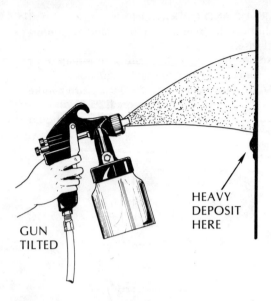

Fig. 4-16: A tilted gun can cause an uneven pattern, run-over and streaking.

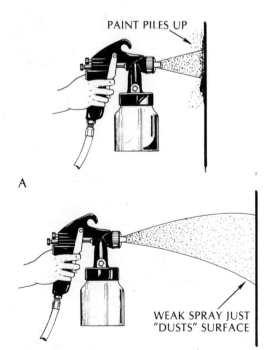

Fig. 4-17: The correct gun-to-work distance is important. If the gun is too close (A) the finish material piles up and causes runs and sags. When the gun is too far away (B) material tends to dry into dust before it reaches the surface. Adjust distance accordingly.

the stroke beyond the work edge (Fig. 4-18). From this position you are ready to go into the next stroke.

Lap each stroke over the preceding one by about 50 percent on each pass. In other words, the aiming point for each stroke is the bottom of the preceding stroke, as shown in Fig. 4-19. This system gives double coverage and assures a full wet coat without streaks. The full half-lap is especially good for lacquer. It is sometimes practical to use less lap, but until you are experienced, it is best to use the half-over-

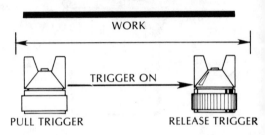

Fig. 4-18: Press trigger at the beginning of the stroke, release at the end with slight overspray, and move gun evenly.

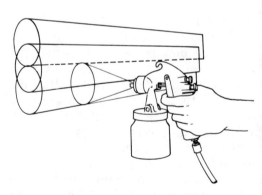

Fig. 4-19: Overlap the strokes by one-half.

lapping stroke for all work.

In order to reduce overspray and to assure full coverage, many finishers use a banding technique (Fig. 4-20). The single vertical stroke at each end assures complete coverage and eliminates the waste of material that results from trying to spray to the very edge with the usual horizontal strokes. At the top and bottom of the pan-

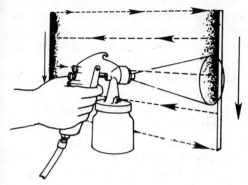

Fig. 4-20: Band spraying the ends.

el, the stroke is aimed at the edge and is an automatic banding stroke. Do not skimp the surface near the edges.

A long panel can be sprayed with vertical strokes (Fig. 4-21). This is sometimes the best system, since stroke-end laps are avoided. However, most sprayers have better control with the more natural horizontal strokes (Fig. 4-22). With a spray gun it is easy to make perfect, invisible overlaps. Spraying surfaces that require overlapping involve the same triggering technique as in spraying a small panel, the only difference being that each area is sprayed as if it were a separate panel. But be sure to spray rather rapidly so that the strokes join. Also, do not try to make strokes which are too long for your arm span. This will surely arc the gun and cause an uneven finish coat.

When a panel is to be sprayed on the edges as well as on the face, a modified banding technique is used. One stroke along each edge coats it and bands the panel. The center is sprayed like an ordinary panel. An outside corner, such as the corner of a box, is finished most effectively with the modified banding technique (Fig. 4-23A). When an inside corner is sprayed square-on, the coating may not be uniform (Fig. 4-23B) but the technique is fast and practical. Use this system for all paints and enamels. When you spray furniture with stain or clear top coats, the preferable method is to spray each side separately (Fig. 4-24). After you make the single vertical stroke at the corner, use

Fig. 4-21: A long panel can be sprayed with up-and-down strokes. When you are spraying horizontally, most long work requires overlap.

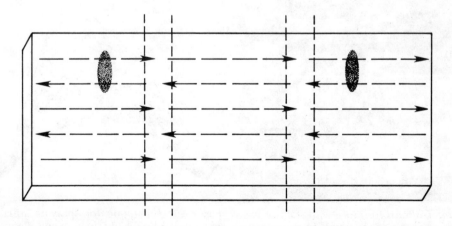

Fig. 4-22: When spraying horizontally, most long work requires overlapping.

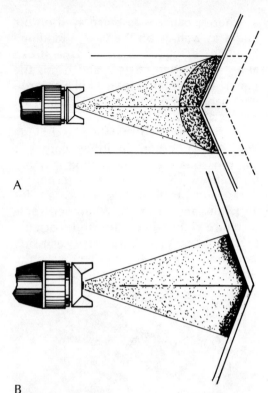

A

B

Fig. 4-23: (A) Outside corner; (B) inside corner. Spraying square-on into a corner results in an uneven coating but is satisfactory for most work.

Fig. 4-24: To finish furniture with stain or varnish, stroke each side separately to give an even coating.

short pull strokes to cover the area adjacent to the corner. The idea here is control—do not overspray or double-coat the adjoining surface.

For slender work, such as table legs, the rule is to make the spray pattern fit the work (Fig. 4-25). Do not use a big horizontal pattern on a slender leg. A smaller horizontal pattern or a big vertical fan gives complete coverage without excessive overspray. On the other hand, do not try to work too small a pattern. The technique of spraying square or rectangular legs is shown in Fig. 4-26.

A flat, round surface is sprayed in the same way as a plane surface, by banding

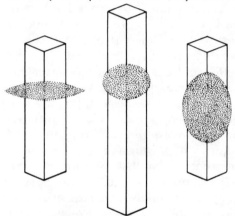

Fig. 4-25: When spraying slender work, avoid excessive overspray by making the spray pattern fit, if possible.

Fig. 4-26: Technique for spraying square or rectangular legs. Hold the gun at a slight angle to avoid overspray on the inside of the legs.

the edge and then spraying the center (Fig. 4-27). Large to medium cylinders are usually sprayed like a flat panel except that the the strokes are shorter. Cylinders can be sprayed with either a vertical or horizontal stroke, as shown in Fig. 4-28. Very small rounds, such as table leg turnings, are sprayed with a vertical stroke, using three or four lapping strokes to get full coverage (Fig. 4-29).

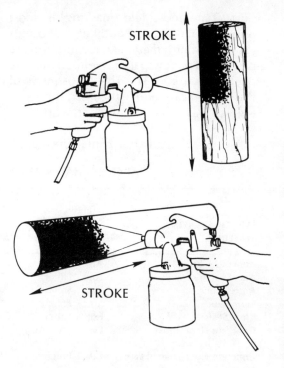

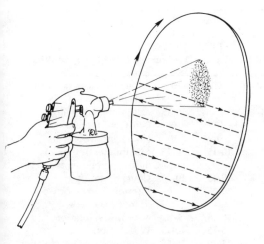

Fig. 4-27: Spraying a flat, round surface.

Fig. 4-28: Round surfaces can be sprayed either vertically or horizontally.

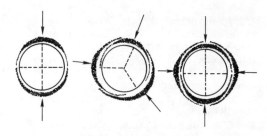

Fig. 4-29: Small rounds can be covered in two to four strokes.

Whenever possible, tilt the work so that the gun can be held level and aimed at an angle of 90 degrees. If this cannot be done, then you have no choice but to tilt the gun, but always try to keep the angle to a minimum. Since tilting the gun results in overspray on the work, always start level surfaces on the near side and work to the far side (Fig. 4-30A). This is a must in lacquer work, since lacquer overspray that lands on wet work will dry sandy. Paint and synthetics are also peculiar in this respect.

Edges (or corners), such as on a table top, can be sprayed by tilting the gun *slightly* so that the air cap is directed at the edge. By doing this, you can cover the edge and form end bands on the top surface in the one stroke (Fig. 4-30B).

Never tilt the gun so far that the finishing material in the canister covers the lid. You can, of course, increase the amount of tilt possible by reducing the amount of fluid in the canister. When using a material tank, the amount of tilt is not a factor.

Masking and Shielding. When doing spray work which is next to a surface that does not require the finish, or when using more than one color, masking or shielding is called for. Simple masking may be done in three ways: (1) using a metal straightedge; (2) with masking tape and a cardboard shield; (3) with masking tape and a skirt. Tape-and-shield masking is most popular. Masking tape outlines the boundary where finish material is to stop, and the cardboard or newspaper shield catches the overspray (Fig. 4-31). The adjustment of the pattern can reduce the need for masking, as was done for slender work (Fig. 4-25).

A vertical fan on a horizontal line is practical but requires a heavy overspray. A horizontal fan gives the least overspray. When you are using a short shield, it is usually simpler to pull separate strokes away from the tape rather than use a single run-in stroke. Cutting-in against a vertical line is a perfect example. The cut-in with the vertical fan gives the least overspray, and, at the same time, the adjoining sur-

PAINTING MATERIALS AND HOW TO SPRAY THEM

TYPE OF SPRAYING MATERIAL	TYPE OF NOZZLE*	GUN†	THINNER‡	PERCENT OF THINNER AT PRESSURES		GENERAL REMARKS
				35 psi	40 psi up	
Aluminum Paint	Any	Either	T-M	0	0	Exterior surfaces, radiators, etc.
Anti-Rust Primer	Any	Either	T	20	10-15	Galvanized metal
Auto Enamel	Ext.	Siphon	M	20	10-15	Automobile spraying
Barn Paint	Any	Pressure	T	10	0	Barns and fences
Clear Sealer	Ext.	Either	T-M	0	0	Floors and woodwork
Concrete Floor-Epoxy	Ext.	Pressure	L	10-15	10-15	Mix thinner at time of mixing components
Concrete Floor-Enamel Latex	Ext.	Pressure	W	10-15	10-15	Concrete floors
Enamel	Any	Either	T-M	10-15	5-10	Interior or exterior wood or metal surfaces
Enamel Undercoat	Int.	Pressure	T	10-15	5-10	New wood, interior
Epoxy Enamel	Ext.	Pressure	L	10-15	10-15	Wood, metal
Epoxy Swimming Pool	Ext.	Pressure	L	15-20	10-15	Swimming pools and patios
House Paint—Oil	Int.	Pressure	T	10-15	5-10	Exterior wood
House Paint Undercoat	Int.	Pressure	T	15-20	10-15	New exterior wood surfaces
Interior Latex	Ext.	Pressure	W	10-15	5-10	Interior walls, ceilings, trim
Lacquer, clear	Ext.	Siphon	L	25-30	20-25	Various surfaces
Lacquer, pigmented	Ext.	Either	L	100	75-100	Various surfaces
Latex—Exterior House	Ext.	Pressure	W	25	15-20	Any exterior surface
Flat Oil Finish	Int.	Pressure	M	10-15	5-10	Interior walls
Red Chromate Primer (Auto)	Ext.	Siphon	M	10-15	5-10	Undercoat for auto
Redwood Finish	Any	Either	M	0	0	Exterior redwood or cedar
Semi-Gloss Enamel—Latex	Ext.	Pressure	W	10-15	5-10	Interior surfaces
Semi-Gloss Enamel—Oil	Any	Either	M	10-15	5-10	Interior surfaces
Shellac—4 lb. cut	Ext.	Either	A	50	40	Floors, furniture, woodwork
Shingle Stain	Any	Either	T	0	0	Exterior wood stain
Stucco, Asbestos Shingle, & Masonry Paint—Latex	Ext.	Pressure	W	25	15-20	Exterior masonry except very porous surfaces
Trim Paint—Oil Exterior	Int.	Pressure	T	10-15	5-10	Exterior wood
Varnish—Interior, Exterior	Int.	Pressure	T-M	10-15	5-10	Exterior wood and metal
Wallpaper Coating, Plastic	Int.	Pressure	M	0	0	Seal wallpaper
Wall Primer and Sealer	Int.	Pressure	T-M	0	0	Seal new walls
Wiping Stains	Ext.	Siphon	M	0	0	Furniture
Wood Stains	Ext.	Siphon	W-M	0	0	Furniture

*Int.—Internal Mix nozzle; Ext.—External Mix nozzle.
†Press.—Pressure-Feed gun.
‡T—Turpentine; M—Mineral Spirits; W—Water; L—Lacquer Thinner; A—Alcohol.
Note: The combination of internal mix-siphon feed will not operate.

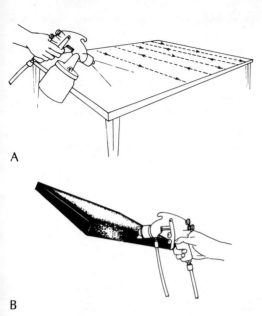

Fig. 4-30: (A) Spraying a level surface; (B) method of finishing a table top.

face is sprayed with the same pattern and the horizontal stroke which most sprayers favor. When using any of these techniques, if the shield is held close to the near edge of the tape, the tape can be used several times. But when using tape, be sure to press down the edges so that a firm contact with the work is established. A good way to do this is with a short piece of dowel.

The taped skirt offers complete protection, making it an ideal device for all exacting spray jobs. Make up the taped skirts to suit the job. For straight line work, prepare the skirting on a clean board.

Masking with a metal shield or a straightedge is the fastest system and is often practical for rough work. The spray should be directed slightly away from the edge. Whatever method is used, cutting-in is cleaner and easier if you use a small spray pattern.

CLEANING EQUIPMENT

Cleaning spraying equipment is an easy but important operation. If done systematically and thoroughly after each job, it will pay dividends in better spraying and trouble-free performance. Cleaning should be done promptly after the spraying job is finished.

Gun Used With a Canister. When the spray gun is used with a canister, the cleaning operation should be carried out as follows:

1. With the compressor turned OFF, unscrew the canister. Pull the trigger and let the material drip into the canister (Fig. 4-32A). Then empty its contents back into the proper container.

2. Rinse the canister with a solvent or thinner appropriate for the spray liquid used (lacquer thinner for lacquer, mineral spirits for oil base paints, warm soap for latex, and so on).

3. Discard the solvent and refill the canister one-half full with fresh solvent. Attach the canister to the gun and spray until the solvent is free of any traces of spray material. While spraying the solvent, shake the gun vigorously. Spray into a cardboard box (Fig. 4-32B).

4. Soak a cloth in the spray solvent and wipe the gun body clean (Fig. 4-32C). *Caution:* Store or discard the solvent-soaked cloth in a metal container that is remote from sparks, flame, or heat.

5. Repeat Steps 2, 3, and 4, using clean solvent.

6. Remove the air cap and the air cap

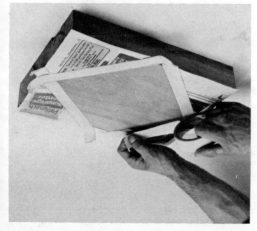

Fig. 4-31: A simple tape-and-shield mask.

seal, and clean them thoroughly in solvent. Use a toothpick or broomstraw to remove the paint lodged in the air cap holes (Fig. 4-32D). Do not use a metal pick; a metal pick may damage the precision-machined surfaces.

7. Wipe the threads clean on both the canister (Fig. 4-32E) and the gun. It may be necessary to use a toothbrush to remove any stubborn material (Fig. 4-32F).

8. Carefully clean the "O" rings (if the gun has them) with a rag soaked in the proper solvent. Avoid soaking any gaskets or "O" rings in solvent as this procedure will reduce the life of such parts. (The gasket and "O" rings frequently act as seals between moving or metal parts carrying air or finish material under pressure. Whenever they lose their life or shape, they should be replaced.)

9. Reassemble the air cap to the gun. Apply a few drops of light household oil to all moving parts and a thin coating of petroleum or mineral type grease to the outside of the canister and to its threads. If a water base solvent has been used, spray the gun with mineral spirits to eliminate residual moisture.

Gun Used With A Pressure Material Tank. When the spray gun is used in conjunction with a pressure material tank, the cleaning procedure is as follows:

1. Shut off the air pressure at the air regulator by turning the valve counterclockwise, and then open the air relief or safety valve so that all of the air trapped in the hose is removed.

2. Completely cover the spray gun nozzle with a soft cloth, and with the compressor running, pull the trigger of the gun. This will force spray material remaining in the hose back into the tank.

3. Remove the tank lid and pour any remaining paint back into its original con-

A

B

C

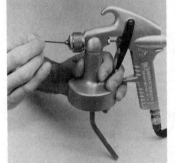
D

E

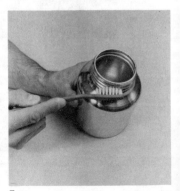
F

Fig. 4-32: Steps in cleaning a canister-equipped spray gun.

tainer so that it may be saved for the future. Clean out the tank and then pour in about one-half gallon of the proper solvent for the type of finish material used.

4. Clamp on the lid, adjust the regulator and compressor for the desired pressure, and proceed to spray the proper solvent through the gun and the hose. Continue until only clear fluid is being sprayed.

5. Again shut off the air to the tank at the air regulator, open the relief or safety valve, cover the gun nozzle, and proceed as indicated in Step 2 to remove all the solvent from the hose.

6. Clean out the tank and carefully dry it (Fig. 4-33). Remove the material hose and air line from the gun and tank. Then clean the air cap and gun parts as described for "Gun Used With A Canister." Store the material tank, gun, and hose in a dry location.

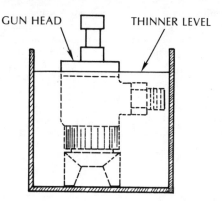

Fig. 4-34: Cleaning the gun head.

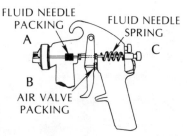

Fig. 4-35: Points that should be lubricated: (A) the fluid needle packing and (B) the air valve packing. A drop or two of oil should be put on the fluid needle packing occasionally to keep it soft. The fluid needle spring (C) should be coated with petrolatum.

Remember that efficient operation of a spray gun depends on proper maintenance and it should be thoroughly clean each time it is used.

Fig. 4-33: Carefully dry the tank after use.

It is very poor practice to place the entire gun in thinner. When this is done the solvent will shorten the life of packings and seals, or mar the surface of the gun body casting. While it is usually not required, it is permissible to place the fluid carrying section of the gun in thinner. Do this by separating the head from the valve control body and placing the head in the thinner (Fig. 4-34). After cleaning, reinstall the head on the valve control body. Figure 4-35 shows the parts of the gun that require lubrication.

Check the owner's manual for any special cleaning or service instructions that apply to your spray gun and air system.

SPRAY GUN PERFORMANCE PROBLEMS

The spray gun, if not adjusted, manipulated, and cleaned properly, will apply a defective coating to the surface. Fortunately, defects from incorrect handling and improper cleaning can be tracked down quite readily, and then corrected without much difficulty. The most common spraying troubles, with their possible causes and suggested remedies, are listed in the general spray gun performance chart below. For specific service information on your spray gun, check the owner's manual.

Trouble	Possible cause	Suggested remedies
Spray pattern top heavy (Fig. 4-36A) or bottom heavy (Fig. 4-36B)	1. Horn holes partially plugged (external mix). 2. Obstruction on top side of fluid tip. 3. Dirt on air-cap seat or fluid-tip seat.	1. Remove air cap and clean. 2. Remove and clean. 3. Remove and clean seat.
Spray pattern heavy to right (Fig. 4-36C) or to left (Fig. 4-36D)	Air cap dirty or orifice partially clogged. To determine where buildup occurs, rotate cap 180 degrees and test spray. If pattern shape stays in same position the condition is caused by material buildup on fluid tip. If pattern changes with cap movement the condition is in the air cap. Clean air cap, orifice and fluid tip accordingly.	
Spray pattern heavy at center (Fig. 4-36E)	1. Atomizing pressure too low. 2. Material of too great viscosity. 3. Fluid pressure too high for air cap's normal capacity (pressure feed).	1. Increase pressure. 2. Thin material with suitable thinner. 3. Reduce fluid pressure.
Spray pattern split (Fig. 4-36F)	1. Not enough material. 2. Air cap or fluid tip dirty.	1. Reduce air pressure or increase fluid flow. 2. Remove and clean.

A B C D E F

Fig. 4-36: Common pattern faults.

Pinholes	1. Gun too close to surface. 2. Fluid pressure too high. 3. Material too heavy.	1. Stroke 6 to 9 inches from surface. 2. Reduce pressure. 3. Thin material with thinner.
Blushing or a whitish coat of lacquer	1. Absorption of moisture. 2. Too quick drying of lacquer.	1. Avoid spraying in damp, humid, or too cool weather. 2. Correct by adding retarder to lacquer.
"Orange peel" (surface looks like orange peel) (Fig. 4-37A)	1. Too high or too low an atomization pressure. 2. Gun too far or too close to work. 3. Material not thinned. 4. Improperly prepared surface. 5. Gun stroke too rapid. 6. Using wrong air cap. 7. Overspray striking a previously sprayed surface. 8. Material not thoroughly dissolved. 9. Drafts (synthetics and lacquers). 10. Humidity too low (synthetics).	1. Correct as needed. 2. Stroke 6 to 9 inches from surface. 3. Use proper thinning process. 4. Surface must be prepared. 5. Take deliberate, slow stroke. 6. Select correct air cap for the material and feed. 7. Select proper spraying procedure. 8. Mix material thoroughly. 9. Eliminate excessive drafts. 10. Raise humidity of room.

Trouble	Possible cause	Suggested remedies
Excessive spray fog or overspray	1. Atomizing air pressure too high or fluid pressure too low. 2. Spraying past surface of the product. 3. Wrong air cap or fluid tip. 4. Gun stroked too far from surface. 5. Material thinned out too much.	1. Correct as needed. 2. Release trigger when gun passes target. 3. Ascertain and use correct combination. 4. Stroke 6 to 9 inches from surface. 5. Add correct amount of thinner.
No control over size of pattern	1. Air cap seal is damaged. 2. Foreign particles are lodged under the seal.	1. Check for damage, replace if necessary. 2. Make sure surface that this sets on is clean.
Sags or runs (Figs. 4-37B and C)	1. Dirty air cap and fluid tip. 2. Gun manipulated too close to surface. 3. Not releasing trigger at end of stroke (when stroke does not go beyond object). 4. Gun manipulated at wrong angle to surface. 5. Material piled on too heavy. 6. Material thinned out too much. 7. Fluid pressure too high. 8. Operation too slow. 9. Improper atomization.	1. Clean cap and fluid tip. 2. Hold the gun 6 to 9 inches from surface. 3. Release trigger after every stroke. 4. Work gun at right angles to surface. 5. Learn to calculate depth of wet film of material. 6. Add correct amount of material by measure. 7. Reduce fluid pressure with material control knob. 8. Speed up movement of gun across surface. 9. Check air and material flow; clean cap and fluid tip.
Streaks	1. Dirty or damaged air cap and/or fluid tip. 2. Not overlapping strokes correctly or sufficiently. 3. Gun moved too fast across surface. 4. Gun held at wrong angle to surface. 5. Gun held too far from surface. 6. Air pressure too high. 7. Split spray. 8. Pattern and material control not adjusted properly.	1. Same as for sags. 2. Follow previous stroke accurately. 3. Take deliberate, slow strokes. 4. Same as for sags. 5. Stroke 6 to 9 inches from surface. 6. Use least air pressure necessary. 7. Reduce air adjustment or change air cap and/or fluid tip. 8. Re-adjust.
Gun sputters constantly (Fig. 4-37D)	1. Connections, fittings and seals loose or missing. 2. Leaky connection on fluid tube or fluid-needle packing (siphon gun). 3. Lack of sufficient material in container. 4. Tipping container at an acute angle.	1. Tighten and/or replace as per owner's manual. 2. Tighten connections; lubricate packing. 3. Refill container with material. 4. If container must be tipped, change position of fluid tube and keep container full of material.

Trouble	Possible cause	Suggested remedies
	5. Obstructed fluid passageway.	5. Remove fluid tip, needle, and fluid tube and clean.
	6. Material too heavy (siphon feed).	6. Thin material.
	7. Clogged air vent in canister top (siphon feed).	7. Clean.
	8. Dirty or damaged coupling nut on canister top (siphon feed).	8. Clean or replace.
	9. Fluid pipe not tightened to pressure-tank lid or pressure-cup cover.	9. Tighten; check for defective threads.
	10. Strainer is clogged up.	10. Clean your strainer.
	11. Packing nut is loose.	11. Make sure packing nut is tight.
	12. Fluid tip is loose.	12. Tighten fluid tip. Torque to manufacturer's specifications.
	13. "O" ring on tip is worn or dirty.	13. Replace "O" ring if necessary.
	14. Material hose from paint tank loose.	14. Tighten.
	15. Jam nut gasket installed improperly or jam nut loose.	15. Inspect and correctly install or tighten nut.
Uneven spray pattern	1. Damaged or clogged air cap.	1. Inspect air cap and clean or replace.
	2. Damaged or clogged fluid tip.	2. Inspect fluid tip and clean or replace.
Material leaks from spray gun (Fig. 4-37E)	1. Fluid-needle packing nut too tight.	1. Loosen nut; lubricate packing.
	2. Fluid-needle packing dry.	2. Lubricate needle and packing frequently.
	3. Foreign particle blocking fluid tip.	3. Remove tip and clean.
	4. Damaged fluid tip or fluid needle.	4. Replace both tip and needle.
	5. Wrong fluid-needle size.	5. Replace fluid needle with correct size for fluid tip being used.
	6. Broken fluid-needle spring.	6. Remove and replace.
Material leaks from packing nut (Fig. 4-37F)	1. Loose packing nut.	1. Tighten packing nut.
	2. Packing is worn out.	2. Replace packing.
	3. Dry packing.	3. Remove and soften packing with a few drops of light oil.
Material leaks through fluid tip when trigger is released	1. Foreign particles lodged in the fluid tip.	1. Clean out tip and strain paint.
	2. Fluid needle has paint stuck on it.	2. Remove all dried paint.
	3. Fluid needle is damaged.	3. Check for damage, replace if necessary.
	4. Fluid tip has been damaged.	4. Check for nicks, replace if necessary.
	5. Spring left off fluid needle.	5. Make sure spring is replaced on needle.
Excessive material	1. Not triggering the gun at each stroke.	1. It should be a habit to release trigger after every stroke.
	2. Gun at wrong angle to surface.	2. Hold gun at right angles to surface.
	3. Gun held too far from surface.	3. Stroke 6 to 9 inches from surface.
	4. Wrong air cap or fluid tip.	4. COrrect combination.
	5. Depositing material film of irregular thickness.	5. Learn to calculate depth of wet film of finish.
	6. Air pressure too high.	6. Use least amount of air necessary.
	7. Fluid pressure too high.	7. Reduce pressure.
	8. Material control knob not adjusted properly.	8. Re-adjust.

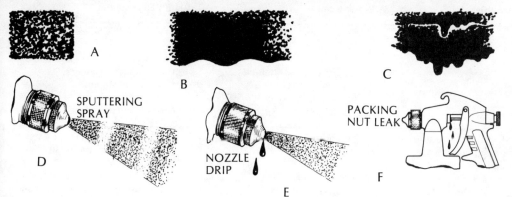

Fig. 4-37: Other spray gun performance problems.

Trouble	Possible cause	Suggested remedies
Material will not come from spray gun	1. Out of material. 2. Grit, dirt, paint skin, etc., blocking air gap, fluid tip, fluid needle, or strainer. 3. No air supply. 4. Internal mix cap using siphon feed.	1. Add more spray material. 2. Clean spray gun thoroughly and strain spray material; always strain material before using it. 3. Check regulator. 4. Change cap or feed.
Material will not come from material tank or canister	1. Lack of proper air pressure in material tank or canister. 2. Air intake opening inside material tank or canister clogged by dried-up finish material. 3. Leaking gasket on material tank cover or canister top. 4. Gun not converted correctly between canister and material tank. 5. Blocked material hose. 6. Connections with regulator not correct.	1. Check for air leaks or leak of air entry; adjust air pressure for sufficient flow. 2. This is a common trouble; clean opening periodically. 3. Replace with new gasket. 4. Correct per owner's manual. 5. Clear. 6. Correct per owner's manual.
Will not spray on pressure feed	1. Control knob on canister cover not open. 2. Canister is not sealing. 3. Spray material has not been strained. 4. Spray material in canister top threads. 5. Gasket in canister top worn or left out. 6. No air supply. 7. Material too thick. 8. Clogged strainer.	1. Set this knob for pressure spraying. 2. Make sure canister is on tightly. 3. Always strain before using. 4. Clean threads and wipe with grease. 5. Inspect and replace if necessary. 6. Check regulator. 7. Thin material with proper thinner. 8. Clean or replace strainer.
Will not spray on siphon feed	1. Spray material is too thick. 2. Internal-mix nozzle used. 3. Spray material has not been strained.	1. Thin material with thinner. 2. Install external-mix nozzle. 3. Always strain before use.

Trouble	Possible cause	Suggested remedies
	4. Hole in canister cover clogged.	4. Make sure this hole is open.
	5. Gasket in canister top worn or left out.	5. Inspect and replace if necessary.
	6. Plug or clogged strainer.	6. Clean or replace strainer.
	7. Material control knob adjusted incorrectly.	7. Correct adjustment.
	8. No air supply.	8. Check regulator.
Air continues to flow through gun when trigger has been released (on non-bleeder guns only)	1. Air control valve leaks.	1. Remove valve, inspect for damage, clean valve, and replace if necessary.
	2. Needle is binding.	2. Clean or straighten needle.
	3. Piston is sticking.	3. Clean piston, check "O" ring and replace if necessary.
	4. Packing nut too tight.	4. Adjust packing nuts.
	5. Control valve spring left out.	5. Make sure you have replaced this spring.
Air leak at canister gasket	1. Canister not sealing on canister cover.	1. Check gasket, clean threads, tighten canister.
Leaks at set screw in canister top	1. Screw not tight.	1. Clean threads and tighten screw.
	2. Damaged threads on set screw.	2. Inspect and replace if necessary.
Leak between top of canister cover and gun body	1. Retainer nut is not tight enough.	1. Check nut to make sure it is tight.
	2. Gasket or gasket seat damaged.	2. Inspect, clean, and replace if necessary.

When using a pressure material tank, there are few problems that may arise, mainly due to air leaks at the top of the lid. If this should occur, it can be corrected as noted here:

Trouble	Possible cause	Suggested remedies
Leaks air at the top of the tank lid	1. Gasket not seating properly, or damaged.	1. Drain off all of the air from material tank thus allowing the gasket to seat. Retighten wing nuts, and fill with air again. Lid will seat tightly.
	2. Wing screws not tight enough.	2. Make sure all wing screws are tight. By following remedy #1 (above), wing screws can be pulled down even tighter.
	3. Fittings leak.	3. Check all fittings and apply pipe dope if necessary.
	4. Air pressure too high.	4. Max. 60 psi. Normal w.p. 25-30 psi.
Regulator not functioning properly	1. See page 55 for regulator troubleshooting procedure.	
No material comes through the spray gun	1. Not enough pressure in tank.	1. Increase regulator setting until material flows; do not exceed 60 psi.
	2. Out of material.	2. Check your material supply.
	3. Material passages clogged.	3. Check tube, fittings, hose, and spray gun. Clean out fittings, hose, tube, and spray gun making sure all residual material is removed.

Chapter 5

OUTDOOR USES FOR AN AIR COMPRESSOR

The air compressor is a great time and money saver when it comes to outdoor work. You will soon find that your compressor, with its capacity for getting more maintenance jobs done faster, will give you more time to enjoy the yard, patio, porch, and other equipment you own. For instance, a compressor with a caulk attachment can plug minor leaks in seams and crevices with no-strain ease and speed. Or a sandblasting gun used in conjunction with a compressor will whisk away rust and remove childish scrawls in no time at all. Do you want to inflate a wading pool? Save your breath (Fig. 5-1). Blow a host of sporting goods up with a compressor and an inflator kit. These are only a few of the many outdoor projects described in this chapter that your compressor will do.

CAULKING

Perhaps the most overlooked form of home insulation is actually the simplest—sealing off air leaks with caulking compound. It can be done in a short time with minimum effort, especially if you use an air-powered caulking gun. While the materials are relatively inexpensive the results can pay big dividends.

Recent United States government (HUD) studies show that caulking air leaks around windows and doors in an average size home can save the owner up to $75 a year in energy costs. As fuel prices climb, so do the savings.

In addition to saving energy, caulking also prevents rotting of wood structures by waterproofing the home. Also caulking is the last step in the preparation work necessary before painting.

Fig. 5-1: Inflating a wading pool is an easy task when using a compressor.

Types of Caulking Compound. There are four basic types of caulking compound available:

1. *Oil-base Caulk.* This economical, general purpose caulk works best on non-moving cracks and joints or to fill joints where little expansion or contraction will occur, such as wooden door frames or cracks in wood. It has a service life of about one year, depending on the manner of application and the climate. Ultimate shrinkage will be about 5 to 20 percent. The life of an oil-base caulk can be extended by frequent painting. Allow it to cure for 24 hours before applying the first coat of paint.

2. *Latex (Water-thinned) Caulk.* Polyvinyl acetate latex-based caulk is generally recommended for interior use. When used for exterior applications, it can wash out if it is rained on before being fully set. Among the latex caulks, polyvinyl acetate is usually lower priced and acrylic latex higher priced. Acrylic latex caulks are easy to apply and can be used on damp surfaces, cleaned up with water, and painted after 30 minutes. Curing occurs by the water evaporating

out of the caulk. It leaves a seal that is flexible and durable. Caulking adheres well to most construction materials, such as wood, concrete, brick, plaster, drywall, plastics, and metal. Latex caulks are available in white and a variety of colors which will not bleed or stain.

3. *Butyl Rubber Caulk.* Butyl rubber caulk is generally used in joints where more than moderate movement may occur, such as metal windows and door frames, gutters, roof flashing, or faucet pipes. It has an excellent ability to stretch, and when applied properly, it will remain flexible in extreme temperatures. Normal shrinkage is 10 to 35 percent. Gun grade butyl caulks become tack-free in 2 to 72 hours, depending on the climate, and they can then be painted. Butyl can be used both above and below grade and has a service life of 7 to 10 years.

4. *Elastomeric Sealants.* Polysulfide and silicone sealants are both considered to be elastomeric compounds, characterized by their rubbery consistency. They are resilient and able to take movement and expansion and contraction in joints without losing their seal. Both silicone and polysulfide caulks have excellent durability and good adhesion to most surfaces. Silicone sealants have a life expectancy of 20 years or more. However, they are more difficult to use and clean up than the less durable caulks. Some may have unpleasant odors, others canot be painted. Polysulfides stay sticky and collect dust and dirt.

All of the above caulks are available in standard 11 ounce paper cartridges, which will fit the typical air-operated caulk gun. One 11 ounce cartridge will caulk 22 to 26 linear feet, if the seam or joint is 1/4 inch deep and wide (see chart here). The largest channel recommended for caulking (1/2 inch deep by 3/4 inch wide) would require one cartridge per 3 feet.

CAULK COVERAGE PER CARTRIDGE
(Approximate linear feet)

Bead depth	Bead width 1/8"	1/4"	3/8"	1/2"	3/4"
1/8"	96	48	32	24	16
1/4"	48	24	16	12	8
3/8"	32	16	11	8	5
1/2"	16	11	8	5	3

Where To Caulk

Caulking should be used wherever two different materials or parts of a house are joined. Figure 5-2 and the following list suggest specific locations to check:

Doors. Check the edges of the threshold inside and out, the ends of the threshold

Fig. 5-2: *Places that need caulking.*

where the door meets the jambs, the joint between the outside door casing and the siding and head casing, sliding glass door tracks.

Windows. Inspect between the casing and siding, underneath windowsills, and on top of the molding over windows.

Siding. Examine any corners where a siding meets a casing; open mitered corners where sidings meet; look for cracks under the bottom course of siding; look between the joints in siding and between the siding and the foundation. To repair damaged siding, pull the split tight and nail it, then force caulk into the crack.

Other Areas. Check at the chimney and roof juncture, between eaves and brick, around exterior outlets, dryer vent and gable vents. Foundation cracks, skylights, joints in concrete steps, porch-to-house joints, stucco cracks, wall-to-slab joints, the space around window mounted air conditioners, and any small gap where air and moisture can pass through are all places to inspect and possibly caulk.

Applying Caulking Compound

There are a few basic steps in applying caulk, which when followed produce the best results. To insure a good adhesion of new caulk, the working surfaces should be free from moisture, paint chips, dirt, grease, and old caulk.

To remove grease, use a rag soaked in mineral spirits. Scrape out the old caulk and loose paint with a putty knife and brush the area with a stiff bristle brush. A sandblasting gun can be used to remove foreign matter from metal and masonry surfaces, while an air-powered disk sander (see page 125) is useful for accomplishing the same purposes on wooden surfaces. Priming of all surfaces is recommended. Steel surfaces should be primed with a metal, rust-inhibitive primer. Check the label directions of the individual products to be sure of their correct use.

The best time to caulk is during the warmer months of spring, summer, and early fall before temperatures turn cold. When applying caulk in temperatures below 40 degrees F, it may not cure properly and the condensation of moisture can prevent a solid bond. If it becomes necessary to caulk at colder temperatures, heat the cartridge in a heating pad or place the material beside the hot water heater before applying it to the joint.

Loading the Air-operated Caulk Gun. To load the air-operated caulk gun (Fig. 5-3), proceed as follows:

1. Cut the end of the caulk cartridge tube spout with a razor knife positioned at a 45 degree angle (Fig. 5-4). The amount cut off will determine the width of the bead of material dispensed, so it is best to cut a small amount off the end and cut more if the bead is not wide enough. (Some spouts are marked with caulking widths such as

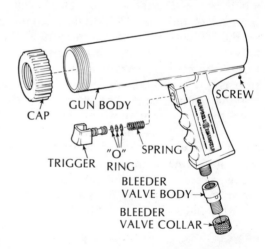

Fig. 5-3: Parts of a typical air-operated caulk gun.

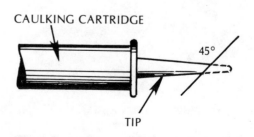

Fig. 5-4: Cut off a bit of the cartridge tip at approximately 45 degrees.

1/4 or 1/2 inch.) The size of the opening should be slightly smaller than the smallest gap to be filled. Insert a wire rod into the tube spout to puncture the inner seal. Be sure the back end of the cartridge is not dented or frayed, or leakage will occur. Also, it is not advisable to use a caulk cartridge if the material inside has dried out. Test it first by pushing a rod through the spout to loosen the material and see if it comes out suitably. If too much pressure is required to loosen the material, discard the cartridge and use a fresh one.

2. Unscrew the front cap and insert the caulk cartridge with the 45 degree angle cut on the spout when facing downwards (Fig. 5-5). The plunger in the back of the cartridge should be free to move. Occasionally, a plunger will flip because of too much air pressure and air and material will come out of the spout. If this occurs, straighten the plunger and operate the gun at a lower pressure. Actually, for the best control of the caulking material, set air pressure as low as possible. A few of the compounds mentioned earlier are protected from moisture contamination by a hermetic seal at the rear end of the cartridge, along with a small pack of moisture absorbing material. When these types of materials are used, it is recommended that the metal cover and the small bag be removed from the tube before inserting it in the gun.

Fig. 5-5: Installing a caulk cartridge in the gun.

3. After the cartridge is in the gun, thread it on the cap until it is hand tight. If the material comes out slowly, it may be necessary to retighten the front cap to seal the caulk cartridge. Regulating the air pressure to the gun is recommended. Most caulk and sealant materials will dispense satisfactorily from 5 psi to 20 psi. *Higher pressure can be used for thick materials but under no circumstances should the pressure be raised higher than 70 psi.*

4. Most caulk guns have a bleeder valve assembly. This device should be threaded tightly onto the thread insert at the bottom of the gun handle and the air hose is then tightened onto the bleeder valve. The bleeder valve will provide a finer adjustment of air flow when the gun is used with regulated air. If the gun is connected to an unregulated air source, the bleeder valve must be used. To adjust the amount of caulk material dispensed from the gun, the knurled ring on the device is rotated until suitable flow is obtained. Adjust the valve for proper flow rate and bead size on a piece of cardboard before the caulk material is actually applied.

5. To dispense the caulking compound, pull the trigger all the way and material will start to flow after the nozzle is filled. Releasing the trigger allows pressurized air to come out of the small hole near the trigger so that the material will not continue to come out of the gun. There may be some flow after releasing the trigger because of the compressible nature of some caulking materials.

Application Techniques. Push the tip of the spout along the joint, applying a smooth, steady pressure on the trigger to form a solid bead. Hold the gun at 45 degrees and parallel to the joint, not at a right angle. Any excess or uneven caulk can be smoothed with a fingertip or small trowel, then wet with water or mineral spirits (check manufacturer's directions) depending on the type of caulk being used. Disengage the trigger to stop the caulk from flowing and remove the gun from the work. Figure 5-6 shows the technique of applying the caulking compound to the various surfaces and portions of a home.

Fig. 5-6: Applying caulking to various parts of the house: (A) apply a bead of caulk around doors and windows where siding meets trim, then tool and smooth material firmly into place; (B) seal cracks between siding and foundation, as well as at siding corner joints and cracks in masonry walls; (C) seal rain gutter joints and where downspout meets gutter, making sure rust and corrosion are removed and area dry; (D) seal around lamp fixtures, TV antenna, utility wire entry, electrical outlets, and plumbing to prevent water seepage; (E) run a bead of caulk down seam between flashing and roof shingles and between flashing and masonry; (F) fill in seams around vent ducts and air conditioning units by applying a bead of caulk between siding and fixture.

To caulk deep joints, back-up filler should be used. If the joint is over 1/2 inch deep, a filler such as polyethylene, polyurethane, and non-staining rubber filler are recommended before caulking. Read the product label for the manufacturer's recommendations.

Caulks should be smoothed after applying them. Smooth the joints flush, rather than concave. A concave surface will develop through natural curing. Before the caulk cures, any excess should be cut and trimmed away. Keep a wiping cloth with solvent within reach to wipe off excess caulking compound. Oil-base and butyl caulks can be cleaned up with mineral spirits; latex can be cleaned up with soap and water. Caulking compounds themselves should not be thinned with oils or thinners because shrinkage or loss of adhesion may occur. Keep the unused portion fresh by covering the tip with masking tape or plugging the spout with a screw or piece of dowel.

EXTERIOR PAINTING

One of the easiest and certainly one of the most effective ways to improve the appearance of your home is by the application of a coat of paint; this is especially true in the case of your home's exterior.

Selection of Exterior Paint

Even though there are a number of different types of paint, selection need not be too much of a problem. First consider the type of surface. Are you painting wood, metal, or masonry? Some paints can be used on all three; others on two. The condition of the surface may also be important. Old chalky surfaces, for example, are not generally a sound base for latex or water-base paints.

Next consider any special requirements. For example, nonchalking paint may be advisable where chalk rundown would discolor adjacent brick or stone surfaces. Or if mildew is a problem in your area, you may use the mildew-resistant paint. Lead-free paints may be used in areas where sulfur fumes cause staining of paints containing lead pigments.

Color is a third consideration, but it is mostly a matter of personal preference. Some colors are more durable than others, and some color combinations are more attractive than others. Your paint dealer can help you with decisions on color durability and combinations.

House paint is the commercial term for exterior paints mixed with many different formulations. It is the most widely used type of paint. Formulations are available for use on all surfaces and for all special requirements such as chalk or mildew resistance. White is the most popular color.

Exterior paint comes in both oil-base and latex (water-base) types. The vehicle of oil-base paint consists usually of linseed oil plus turpentine or mineral spirits as the thinner. Latex paint contains water as the vehicle thinner; its vehicle consists of fine particles of resin emulsified or held in suspension in water.

Another type of water-base paint has a vehicle consisting of a soluble linseed oil dissolved in water. This paint has the properties of both oil-base and water-base paints.

The advantages of latex paints include easier application, faster drying time, usually better color retention, and resistance to alkali and blistering. Latex paints can also be applied in humid weather and to damp surfaces. Spray gun and tool cleanup is simpler because it can be done with water.

Use the chart on page 87 as a guide in selecting paint; your paint dealer can also help you. The following are some specific suggestions:

Wood Surfaces. Wooden clapboard siding, one of the most commonly used exterior building materials, lends itself to almost any house paint formulated for wooden surfaces.

Shingles made of various decorative woods, on the other hand, may have a natural grain which is pleasing to the eye. It can be coated with a clear water-repellent

	House paint (oil or oil-alkyd)	Cement powder paint	Exterior clear finish	Aluminum paint	Wood stain	Roof coating	Trim paint	Porch and deck paint	Primer or undercoater	Metal primer	House paint (latex)	Water-repellent preservative
Masonry												
Asbestos cement	x.								x		x	
Brick	x.	x		x					x		x	x
Cement and cinder block	x.	x		x					x		x	
Concrete/Masonry porches and floors								x			x	
Coal-tar-felt roof						x						
Stucco	x.	x		x					x		x	
Metal												
Aluminum windows	x.			x			x.			x	x.	
Steel windows	x.			x.			x.			x	x.	
Metal roof	x.									x	x.	
Metal siding	x.			x.			x.			x	x.	
Copper surfaces			x									
Galvanized surfaces	x.			x.			x.			x	x.	
Iron surfaces	x.			x.			x.			x	x.	
Wood												
Clapboard	x.			x					x		x.	
Natural wood siding and trim			x		x							
Shutters and other trim	x.						x.		x		x.	
Wood frame windows	x.			x			x.		x		x.	
Wood porch floor								x				
Wood shingle roof					x							x

Note: The dot in X. indicates that a primer sealer, or fill coat, may be necessary before the finishing coat (unless the surface has been previously finished).

preservative. However, color can be added by applying two coats of a quality pigmented stain, which will enhance the beauty of the shingles, seal the surface, and provide protection against the weather.

Wooden trim, such as window sashes, shutters, and doors, should be attractively coated with a colorful exterior enamel. These coatings, which dry with a relatively glossy surface, are available in either water- or oil-based mixtures, and in a variety of sheens. Those which have the smoothest surface are called high gloss enamels, while others are classified as semi-gloss coatings.

Masonry Surfaces. Masonry surfaces—brick, cement, stucco, cinder block, or asbestos cement—can be revamped with a variety of paint products. One of the newest ideas in painting brick is a clear coating that withstands weather and yet allows the natural appearance of the surface to show through.

Cement-based paints are also used on masonry surfaces. Colorful rubber-based coatings, vinyl, and alkyd emulsion paints, are also used on masonry. Almost all exterior house paints may be applied to masonry; however, they are successfully used only when surface preparations are made properly.

Asphalt shingle siding, on the other hand, requires a rather special treatment calling for exterior emulsions formulated for these surfaces.

Metal Surfaces. Galvanized iron, tin, or steel building materials are available in various types, all of which may rust if not protected against moisture. Copper building materials, although they will not rust, will give off a corrosive wash that will discolor surrounding areas. Aluminum, like copper, will not rust but will corrode if not protected.

Conventional house paints or exterior enamels can be applied to these surfaces. But there are some rust-inhibiting coatings that would be better suited for the job. Ask your paint dealer which paint is formulated for application to the metal used on the exterior of your home.

Porches, Decks, and Steps. Porch floors and steps are usually constructed of wood or concrete, factors that should be kept in mind when choosing paint. The most important point to remember, however, is that foot traffic on these areas is extremely heavy, so the paint used must be durable. Most paint stores stock special porch and deck paints that can wear well under this hard use, but the selection of a primer coat will vary according to the building material used. Wooden porches and steps, for instance, can be primed with a thinned version of the top coat, while cement areas may need to be primed with an alkali-resistant primer.

Best results in painting concrete porches and steps can be obtained with a rubber-based coating or similar product. Roughening the surface slightly with muriatic acid is recommended before painting concrete that is hard and glossy, but in any case all instructions on the label should be closely followed.

Estimating the Paint Quantity

For a rough estimate, you can figure the amount of paint you need for flat surfaces by simply multiplying the height (or length) times the width of the surface and dividing the result into the coverage estimate on the paint can label. That is, figure the siding area below the roof line by adding the length of your house to the width, multiplying by the height, and multiplying that number by two. For example (Fig. 5-7), the square footage of a house 40 feet long and 20 feet wide would be (20 + 40) x 12 x 2 = 1440 square feet. To estimate the paint needed for pitched roofs and gables, multiply the height of the peak from the roof

Fig. 5-7: How to estimate the amount of paint needed for a house.

base by half the width of the area, doing this for each peaked area; then add the area of each gable to the below-roof-line siding area. In our example, this is 4 x 10 x 2 = 80 square feet; the total area to be painted thus is 1440 + 80 = 1520 square feet.

If the label states that paint coverage is approximately 400 square feet per gallon, four gallons will be adequate for one coat. Second coats generally require less paint. But remember that these are average estimates of coverage and because the material is thinned slightly more than when brushing, coverage with a spray gun is about 10 percent more than stated on the label. Do not try to spray a product unless it is formulated for spraying, and do not thin it unless directed to. Your paint dealer can give you the benefit of his experience in tailoring an estimate to the particular surfaces you are going to be covering. Be ready to provide him with exact measurements of the areas to be covered.

Your dealer will also be able to advise you about the number of coats that will be required for different surfaces and different types of paints. Do not make any purchases until you and the dealer have reached agreement on the approximate amount of paint, thinner, primer, and other supplies you may need. If a particular dealer is lacking in patience in reaching this agreement, find another dealer. Also, it pays to buy quality products, and this is especially true with paint. The actual cost of paint to cover the average house is very small when considering the total cost of labor and time. Quality pays because of the lasting characteristics of good paint. There is more time between repaintings, with the result that labor costs are reduced. On new wood, apply two finish coats after the primer. On previously painted work, apply one or two coats, depending on the condition of the old paint. Apply two coats if the colors are changed.

Surface Preparation

The finest paint, applied with the greatest skill, will not produce a satisfactory finish unless the surface has been properly prepared. The basic principles are simple, and with the help of air-driven tools, the work is greatly reduced. But the procedures vary somewhat with different surfaces and, to some extent, with different paints, but the goal is the same: to provide a surface with which the paint can make a strong, permanent bond. In general, these principles include the following:

1. The surface must be clean, smooth, and free from loose particles such as dust or old paint. Use an air-driven sander (see page 90), a wire brush, or a scraper to clean the surface.

2. Oil and grease should be removed by wiping the surface to be painted with mineral spirits. If a detergent is used, it should be followed by a thorough rinse with clean water. Your sandblasting gun (Fig. 5-8) does a good job by permitting pressurized washing. The pressure gun described on page 22 also does a fine job in cleaning a surface before spraying (Fig. 5-9). But remember,

Fig. 5-8: The sandblasting gun can be used as a pressure wash.

Fig. 5-9: Cleaning siding with a pressure gun.

when using a detergent to clean a surface, be sure that it has been rinsed off thoroughly with clean, clear water before applying the paint. If you do not rinse the surface, the new paint will not properly adhere.

3. Chipped or blistered paint should be removed with an air-driven sander, a wire brush, steel wool, or a scraper. A sandblast gun can be used on textured-type wood.

4. Chalked or powdered paint should be removed by power washing with the sandblasting gun using water mixed with TSP (trisodium phosphate, sold in hardware stores). If the old surface is only moderately chalked and the surface is relatively firm, an oil primer can be applied without the prior washing. The primer rebinds the loose particles and provides a solid base for the paint.

5. Loose, cracked, or shrunken putty or caulk should be removed by scraping.

6. If new sealants are used, they should be applied to a clean surface as previously described in this chapter, and allowed to harden before the paint is applied. If the caulk is a latex type, latex paint can be applied over it immediately without waiting for the caulk to harden.

7. Damp surfaces must be dry before paint is applied, unless you are using a latex paint.

Wooden Surfaces. Wooden siding and other exterior wooden surfaces preferably should not contain knots or sappy streaks. But if new siding or wood does contain imperfections, clean the knots and streaks with turpentine and seal them with a good knot sealer, such as shellac. The knot sealer will seal in the oily extractives and prevent the staining and cracking of the paint in the knot area. Smooth any rough spots in the wood with the air-driven sander (Fig. 5-10). Fill cracks, joints, crevices, and nail holes with glazing compound, putty, or plastic wood and sand them lightly until flush with the wood. Always sand in the direction of the grain, never across it. Dust the surface just before you paint it.

Fig. 5-10: An air-driven sander can be used on wood siding.

New wooden surfaces to be stain-finished should first be sanded smooth. Open-grain (porous) wood should be given a coat of paste filler before the stain is applied. (Paste fillers come in various colors to match the wood.) The surface should then be resanded. Read the manufacturer's instructions carefully before applying paste fillers.

Old surfaces in good condition, just slightly faded, dirty, or chalky, may need only dusting before being repainted. Very dirty surfaces should be washed with a mild synthetic detergent or TSP and rinsed thoroughly with water. Grease or other oily matter may be removed by cleaning the surface with mineral spirits. Power spray the surface thoroughly and allow it to dry.

Remove all nail rust marks. The nailhead should be set below the surface, primed and the holes puttied. Fasten any loose siding with nonrusting nails. Fill all the cracks; compounds for this purpose are available from paint and hardware stores. Sand the area smooth after the compound dries, removing all rough, loose, flaking, and blistering paint. Spot-prime the bare spot before repainting. Where the cracking or blistering of the old paint extends over large area, remove all the old paint down to bare wood. Prime and repaint the old surface as you would a new wooden surface. Sand or feather the edges of the sound paint before you repaint.

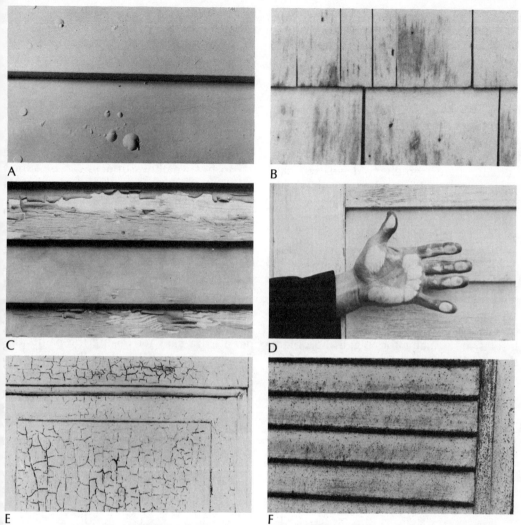

Fig. 5-11: Common paint faults: (left to right, top to bottom) blistering, bleeding, peeling, powdering, checking, and mildew.

Old paint may be removed by sanding, scraping, or with chemical paint remover. Scraping is the simplest but hardest method. Sanding is most effective on smooth surfaces. Chemical paint remover can be expensive for large areas. Remember that it is of utmost importance to correct the condition that caused the blistering, cracking, or peeling of the old paint, before you repaint; otherwise, you may run into the same trouble again (Fig. 5-11).

Masonry Surfaces. New concrete should weather for several months before being painted. If earlier painting is necessary, first wash the surface with a solvent such as mineral spirits to remove oil or grease used for hardening the concrete during the "curing" process. Fresh concrete may contain considerable moisture and alkali, so it is probably best to paint it with latex paints.

Patch any cracks or other defects in masonry surfaces, paying particular attention to the mortar joints. Clean both new and old surfaces thoroughly before painting and remove dirt, loose particles, efflorescence (the crystalline deposit that appears on the mortar between the bricks in a brick wall), and loose, peeling, or heavily chalked paint by sandblasting. Oil and grease may be removed by washing the surface with a commercial cleaner or with a detergent and water.

Metal Surfaces. The techniques described on page 98 hold good for metal house surfaces.

When to Spray Paint

You can easily ruin your paint job if you forget to consider the weather. Excessive humidity or extremely cold weather can cause paint troubles. Latex or water-base paints allow more freedom in application than do oil-base paints; the former can be applied in humid weather.

1. Spray paint when the weather is clear and dry and the temperature is between 50 and 90 degrees F. Never paint when the temperature is below 40 degrees F.

2. Do not paint in *windy* or dusty weather or when insects may get caught in the paint. Insects are usually the biggest problem during fall evenings. Do not try to remove insects from wet paint; brush them off after the paint dries.

3. Start spraying after the morning dew or frost has evaporated and stop in the late afternoon or early evening on cool fall days. If the siding has been thoroughly soaked wet by rain, let it dry several days before applying paint. However, when latex paint is used, as previously stated, some moisture can be left on the surface.

4. In hot weather, surfaces can be spray painted after they have been exposed to the sun and are in the shade. That is, follow the sun around your home; never precede the sun (Fig. 5-12). Too much heat will result in an improperly cured finish, meaning shorter life or later paint trouble.

Application of Exterior Paint

Before starting any exterior paint job, be sure that all the tools and materials are at hand. Along with the proper paint, you will need extra cans for mixing the paint and paddles for stirring it. You will need strain-

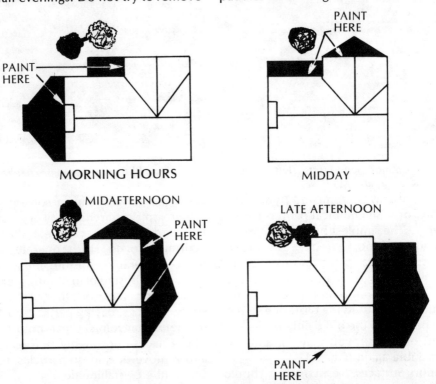

Fig. 5-12: In warm or hot weather, follow the sun around the house so that you are always spray painting in the shade.

ers for removing lumps that might mar the smooth finish and rags for cleaning up spatters. If the house trim is to be a different color from the body, mask off all of the trim. If they are to be the same color, mask only the window glass. Apply 2 inch masking tape around the edges of the glass, leaving the inside of the tape loose. Then using newspaper (Fig. 5-13) or polyethylene sheeting cut into pieces of 1 mil (Fig. 5-14), slip the covering material under the edges of the masking tape and press.

Fig. 5-13: Covering a window with newspaper and masking tape.

Fig. 5-14: This multi-pane picture window was completely masked using tape and polyethylene sheeting.

If you use polyethylene sheeting as a masking material, it can also be employed as a drop cloth to cover shrubs, patios, driveways, walks, or any surface near the house which you do not want to paint. One mil poly is available in 200 foot rolls at your local lumber or home center dealer.

Survey the house to see what the height requirements are. If the house is typical, you will most likely need a 6 foot stepladder, a 12 to 16 foot extension ladder, and a steel scaffolding system. The scaffold has several advantages over a ladder. You will be able to cover about 10 running feet of surface at one time counting the distance of over-reach on either end. You do not have the ladder between you and the work surface. Instead of standing on a ladder rung, you stand on a work platform of planking, which is both safer and more comfortable. The scaffold and planking are usually available at rental outlets. Remember that two sections, 8 feet long, will give nearly 20 running feet of coverage, or the two sections can be stacked easily to reach 18 foot heights.

Though poorly placed scaffolding is more forgiving than a ladder, it is still best to use scraps of wood to block up the legs so they are level. There are also safety railings available to prevent you from falling when stepping back to admire your work. The scaffolding is easy to erect. The X braces are hinged in the middle, and have holes at each end. You simply slip a lock-collar up, slide the hole on the brace over the pin on the ends, and drop the lock collar. There are extra pipe connectors which fit in the top of the ends to allow you to stack the sections for more height. The planking has curved hooks on each end which fit over the tops of the railing. You may find it an advantage to scaffold one side of the house, prepare the surface, and paint that portion before moving the scaffold. This keeps scaffolding assembly time to a minimum. For most exterior house painting jobs, it is a good idea to use a spray gun that is hooked up to a material tank.

Check the label before you mix or stir the paint. Some manufacturers do not recommend mixing, as it may introduce air bubbles. If mixing is required, it can be done at the paint store by placing the can in a mechanical agitator, or you can do it at home with a paddle or spatula (Fig. 5-15). If

Fig. 5-15: Even though your paint dealer has mixed the paint mechanically, mix it again if recommended just before painting. Stir the contents of the can from the bottom up (left); "box" the paint by pouring it from one can to another (right). Then the paint may be poured into the material tank or spray gun canister.

you open the can and find that the pigment has settled, use a clean paddle or spatula and gradually work the pigment up from the bottom of the can, using a circular stirring motion. Continue until the pigment is thoroughly and evenly distributed, showing no signs of color separation. Of course, the job of mixing paint can be made a great deal easier with an air-driven drill equipped with a paint blender or stirrer attachment (Fig. 5-16). They do a first-rate job if you always insert the blender into the can before you start the drill and always stop the drill before you remove the stirrer. Actually, the best way to eliminate spills is to drill a 1/4 inch hole in the can lid to allow blending with the container closed.

Fig. 5-16: An air-driven drill, plus a paint blender, makes the job of mixing paint much easier.

Exterior Paint Spraying Tips

The same basic techniques that are described in Chapter 4 hold true when painting the exterior of a house. The type of air cap and feeding style chosen on the spray gun should be appropriate for the type of material being sprayed (see chart on page 72). As a general rule, it is best to start at a high point of the house (Fig. 5-17), at a corner or under the eave. Actually, the gutters should be sprayed first (Fig. 5-18), then spray from the top to the bottom and then begin again at the top. Treat each of these sidewall sections as you would separate panels, blending them together with lapped strokes as described on page 69. Avoid too thick an application causing paint to run or sag. In fact, it is a good idea to keep a brush or paint pad handy as you spray to wipe away any heavy paint build-ups or sags. If the paint flow is adjusted properly, you will not get many sags on rough surfaces. But be careful on smooth surfaces such as garage doors or plywood soffits. These smooth surfaces should be fogged very lightly with two coats rather than trying to get full coverage with one pass.

When painting siding, first spray the bottom edges, then the flat portion to be sure you are getting full coverage. Pull the trigger as you begin your pass and release it at

Fig. 5-17: Always start at the high point of the house and work down.

Fig. 5-18: Spray paint the gutters first.

the end of the pass. This will help you avoid a heavy overlap and possible running of the paint. As you spray paint, a portable shield (Fig. 5-19) such as a piece of cardboard can be used wherever paint should be kept off of other parts of the structure.

When painting exterior doors, start by spraying the top panels. Spray paint the molding edges first. Then do the remaining panel area with horizontal strokes, then up and down. After doing all the panels, spray the remaining area and finish with the door edges. If the door swings out, paint the lock-side edge with exterior paint. If it swings in, paint the hinged edge with exterior paint. Spray paint flush doors the same way you would a wall or other flat surface, spraying the edges first and then filling in the large area. Complete the job by painting the door frame and jambs.

Windows, masked as shown in Fig. 5-20, can easily be sprayed, but the parts should be painted in the following sequence:

1. Mullions (strips on the glass).
2. Horizontals of the sash.
3. Verticals of the sash.
4. Verticals of the frame.
5. Horizontals of the frame and sill.

When finishing a window, be sure to paint the edges of the casings and do the underside of the sills. While painting, leave the windows slightly ajar at the top and bottom to keep them free; open and close them several times a day until the paint is thoroughly dry.

Complete one sidewall before starting another. In fact, finish a complete side, or at least up to a door or window, before stopping for the day. It is most important that you do not start a new can of paint in the middle of a board or large wall area. If

Fig. 5-19: Using a shield to protect surfaces.

Fig. 5-20: Spraying a window.

the remaining paint in a can will not finish an area, mix some of the new paint with the partially-filled can before starting the area. This will help blend the color.

Because shingles and shakes have many exposed edges, they are susceptible to the absorption of moisture. Moisture may also enter the wood from the back of the shingle. For this reason, the best results are usually obtained by coating the shingles with specially formulated shingle stains, which allow moisture to escape without causing blistering and cracking of the coating film. Sometimes a thicker coating is desired than can be obtained with a stain. This can best be accomplished with the use of exterior latex paint. These paints deposit films that are more porous than high-gloss oil-base paints and are more likely to allow the escape of moisture without blistering, cracking, or peeling. The application of high-gloss coatings to rough surfaces such as shingles and shakes is not generally desirable from an appearance standpoint, in addition to the reasons given above. Wood staining can usually be avoided by sealing the surface with a properly formulated solvent primer or a latex primer especially designed for resistance to staining. It is strongly suggested that the manufacturer's recommendations be followed carefully in selecting and applying a coating system for wooden shingles and shakes.

Porch and deck floor areas constructed of wood, cement, and so on can be finished in minutes, once the surface preparation is complete. Follow the directions for surface preparation usually described in detail on the paint label. Use newspaper to mask the walls by taping at the floor line (Fig. 5-21). Two light coats of varnish or other clear materials should be sprayed on wooden floors, while colored floor, porch, or basement paint will usually cover in one application. Notice that a paint tank is used. It eliminates paint dripping from the canister or gun when it is tilted.

Shutters (Fig. 5-22) should be taken down and sprayed with trim paint. A pair of saw

Fig. 5-21: Spraying a wooden deck.

Fig. 5-22: Spraying shutters.

horses or other supports will make this a a fast, easy job. Like any other surface exposed to the weather, you may find they require sanding and wood-filler in order to achieve a smooth, continuous painting surface. Although only one side is exposed to weather, they should be finished on both sides to prevent rotting and warping.

Storm sashes and screens are troublesome to paint with a brush, but are simple to spray paint (Fig. 5-23). Newspaper shields taped to the glass of a storm sash are quick to apply and can be left on until all the coats are dry. Use a moveable shield of thin sheet metal or cardboard to catch any overspray. Spray the edges of the sash frames head-on first. When the paint is dry, simply peel off the masking. Screens are easy to paint if you spray the screen cloth and the frame the same color (Fig. 5-24). Stack screens behind one another to save paint. Wash them thoroughly by using

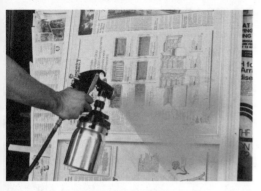

Fig. 5-23: Spraying storm sash.

Fig. 5-24: Spraying screens.

pressurized water from your sandblasting equipment. Spray the screens with screen enamel, exterior spar varnish, or exterior enamel thinned with turpentine. Separate screens after spraying screen cloth and spray paint frames one at a time.

Other Exterior Painting Jobs

There are dozens of painting jobs around the yard that you can finish in short order with a spray gun. For instance, fences around the property (Fig. 5-25) can be spray painted at the same time as you paint the house. To save paint, spray at a slight angle, using a vertical motion on a picket fence. Place a cardboard or other shield behind each section of a chain link fence when spraying.

An outdoor wooden bench or picnic table (Fig. 5-26) should be given regular care to avoid rotting, warping, or cracking of the top and the seats. Usually, such furniture can be sprayed immediately after

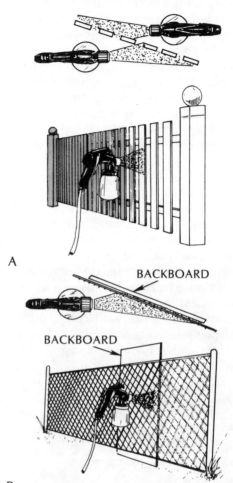

Fig. 5-25: (A) Spray a picket fence at an angle, (B) while a cardboard shield is used behind a chain link fence.

Fig. 5-26: Power cleaning of a picnic table before spraying, using a sandblasting gun.

brushing off the dust and dirt. If the paint has peeled or cracked, remove it completely with a paint and varnish remover or an air-driven sander. Most of these woods require an initial filler, or at least one coat of shellac if an outdoor varnish is to be used later. If an outdoor enamel or trim paint is used, then a primer coat is recommended, applied lightly.

Porch or patio furniture, including wicker, bamboo, or rattan (Fig. 5-27) can be painted to look like new again along with other yard and garden equipment such as rakes, shovels, and lawn rollers. Many of these items are almost impossible to paint with a brush, but they can be sprayed very quickly. Use a wire-brush or the sandblast gun on metal furniture to remove rust and dirt and a stiff bristle brush on wicker or rattan. When painting the latter, thin the paint about 15 percent to avoid build-up of paint in joints of chairs, and apply two coats. Other furniture spraying techniques are given in Chapter 6.

Fig. 5-27: Painting outdoor wicker furniture is easy with a spray gun.

Outdoor trellis or rose arbors can be painted as they stand in the early spring or late fall. When foliage appears, you will also find your sprayer a convenient tool for applying insecticides or weed-killers. Be sure to clean all your spray equipment thoroughly between jobs as weed-killer can also kill flowers and shrubs.

COMBATTING RUST

Rust damage can be costly if left uncontrolled. Garden tools and metal equipment such as outdoor grills, gym sets, tractors, and bicycles are susceptible to rust. Once rust forms, it will continue to eat at the metal until the item is completely ruined. The only way to stop rust is to clean it completely off of the object and then cover it with paint or some other protective material. The cleaning can be done three ways: hand, power tool removal, and abrasive blasting.

Hand Cleaning. Hand cleaning with a wire brush and emery cloth will remove only loose or loosely adhering surface contaminants. These include rust scale, loose rust, mill scale and loosely adhering paint. Hand cleaning is not to be considered an appropriate procedure for removing tight mill scale and all traces of rust. In general terms, hand cleaning cannot be expected to do more than remove major surface contamination. As such, it is primarily recommended for spot-cleaning in areas where corrosion is not a serious factor.

Before hand cleaning, the surface must be free of oil, grease, dirt, and chemicals. This can best be accomplished with solvent cleaners. Then remove rust scale and heavy build-up of old coatings with wire brushes and scrapers. Finish up by sanding, especially on woodwork. All work must be done to avoid deep marking or scratches on the surface by the tools used. Start painting as soon as possible after cleaning.

Power Tool Cleaning. Power tool cleaning methods provide faster and more adequate surface preparation than hand tool methods. Power tools are used for removing small amounts of tightly adhering contaminants which hand tools cannot remove. Power tools still remain time-consuming as compared with blasting for large area removal of tight mill scale, rust, or old coatings. There are several pneumatically-operated tools that do a good job on rust, including needle and rotary

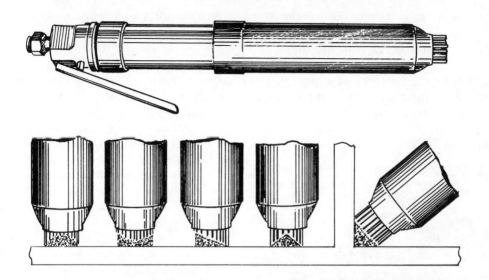

Fig. 5-28: Needle scaler operations.

scalers, grinders, chisels, sanders, and wire brushes. You must be especially careful when using needle and rotary scalers since they will chew up anything in their path. Avoid getting the power line or any part of your body in their way.

Needle scalers accomplish their task with an assembly of individual needles impacting on a surface hundreds of times a minute. The advantage of using individual needles is that irregular surfaces can be cleaned readily. See the operations and how the needle scaler self-adjusts to the contour of various surfaces in Fig. 5-28.

The rotary scaling and chipping tool, sometimes called a jitterbug, has a bundle of cutters or chippers for scaling or chipping (Fig. 5-29). In use, the tool is pushed along the surface to be scaled and the rotating chippers do the work. Replacement bundles of cutters are available when the old ones are worn.

Wire brushes (cups or radial) are available for use with air-driven drills and are used for removing loose mill scale, old paint, slag, and dirt deposits. Grinders and sanders are used for complete removal of old paint, rust, or mill scale on small surfaces, and for smoothing rough surfaces.

Fig. 5-29: A typical rotary impact scaler.

When using chisels and grinding tools, care must be exercised not to cut too deeply into the surface, since this may result in burrs that are difficult to protect satisfactorily. Care must also be taken when using wire brushes to avoid polishing metal surfaces, and thus prevent adequate adhesion of the subsequent coatings. Power tool cleaning is to be preceded by solvent or chemical treatment, and painting must be started and completed as soon after power cleaning as possible.

Blast Cleaning. Blast cleaning abrades and cleans through the high velocity impact of sand, glass beads, synthetic grit, or other abrasive particles hitting the surface. (See the chart on page 20 for the most common abrasive materials, how to use them, and where they are available.) Blast cleaning is most often used on metal but may also be used, with caution, on wooden and masonry surfaces. It is, by far, the most thorough of all mechanical treatments to remove rust (Fig. 5-30).

A

B

Fig. 5-30: (A) Blast cleaning is the best and easiest way to clean metal; (B) it can also be done on masonry surfaces.

When using a sandblast gun, observe the following simple safety rules to insure satisfactory results and to prevent injury:

1. Eye protection and respirator safety equipment should be worn if the eyes and face are exposed to flying particles during operation.

2. Do not expose the hands or skin directly in the line of the blast nozzle.

3. Do not subject the canister to a pressure in excess of 50 psi when using the pressure feed method. By using a proper nozzle in the hopper siphon feed operation, up to 90 psi can be employed. Keep in mind that as the psi, sand flow, and cfm increases, so does work performance.

It is preferred that your sandblasting equipment be used in conjunction with commonly available sandblasting cabinets. However, if such cabinets are not available, then the following tips may be of help. The sandblasting equipment should be used in a well-lighted and well-ventilated area such as a garage, outdoor patio or yard (Fig. 5-31). The chosen area should be sufficiently clear to any objects which could be damaged by the flying abrasives. To localize abrasive scattering or if reclamation of

Fig. 5-31: Cleaning a grill on site with a sandblast gun.

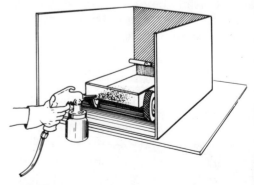

Fig. 5-32: A simple sandblasting enclosure.

the abrasives is desired, some sort of wooden or cardboard enclosure, such as illustrated in Fig. 5-32, may be helpful.

As mentioned in Chapter 2, there are two basic types of sandblasting guns: bleeder (Fig. 5-33) and non-bleeder (Fig. 5-34). The test and problem-solving procedure for a typical sandblast gun of each type is basically the same. That is, disconnect the gun from the canister or hopper hose, and check to determine if there is any siphoning action. Place your finger on

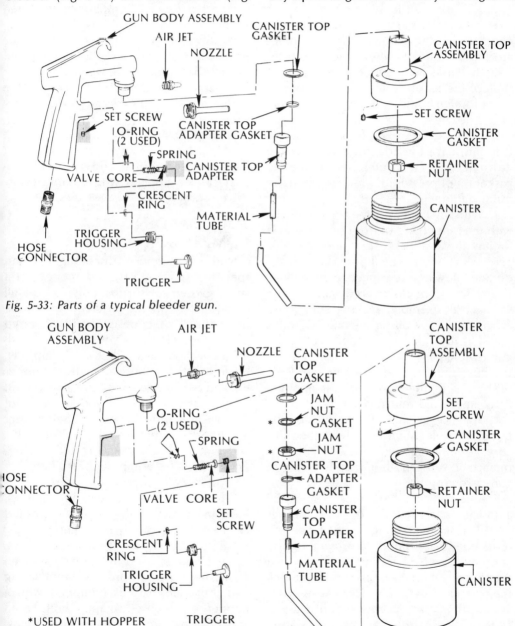

Fig. 5-33: Parts of a typical bleeder gun.

Fig. 5-34: Parts of a typical non-bleeder gun. Major differences in the guns are indicated by the shaded areas.

the bottom of the material tube and pull the trigger. If you do not feel a sucking or siphoning action, then proceed as follows:

1. Take the nozzle off and check it to see if the air jet is in the gun. Make sure the air jet is in tight, and not stripped or cracked. The hole in the air jet must be the correct size. If the air jet is in the gun body properly, check inside the gun body for any holes around the threads where the tube is installed. If the air jet hole is malformed, replace the air jet. Also, check for dirt or other foreign bodies built up on the inside of the nozzle.

2. Examine the gaskets for leaks. If a leak is found, have the adaptor nut tightened down to correct this defect. Make sure the gasket is not pressed out too far when the nut is tightened.

3. With the finger *off* the trigger, check with the other hand in front of the nozzle for any air valve leaks. Try pulling the trigger back a few times by slipping your finger off and allowing the trigger to snap back. If any leaks are discovered, change the valve core assembly, making sure there are not any burrs under the two "O" rings and that they are not cracked. Retest the gun. If leaks are still found, take parts off the gun body so that you can see if the bushing is in the gun body far enough. Reassemble and retest.

4. The trigger should have air coming out of the hole. If it is not, check the valve core assembly. If defective or damaged, replace it with another bleeder type of core assembly.

Frequently, metal surfaces cleaned by abrasive action need no other primer finish prior to painting them. However, remember that primers do provide a good adhesive base for painting metal and that they usually contain inhibitors that block the chemical action which causes rust. Some primers are porous when dry and need to have a topcoat applied as soon as possible to protect them from moisture. Others form a water-resistant shield which can be topcoated at your leisure. If the label does not tell you how long the primer must set before applying the topcoat, be sure to ask your paint dealer.

When choosing a primer, read the label carefully. Look carefully for the words "wood or metal" on the can indicating its appropriate use. Compatibility between primer and topcoat paint is also essential. The solvents in a lacquer topcoat, for instance, may dissolve the same metal primer that would work well with another topcoat. Some primers are compatible with their topcoats because they share similar ingredients.

Suitable paints for commonly-found metal surfaces are listed in the chart on page 87. Both primer and the paint topcoat can be applied with a compressor-driven spray gun.

YOUR GARDEN AND A COMPRESSOR

Nothing is more discouraging than spending time, effort, and money on a lawn, garden, or shrubs, only to have all or part fall before the onslaught of a horde of ravenous insect pests. The hardhearted little beasts chew up your greenery and, if unchecked, may destroy it entirely, blade by blade. But head high, stiff upper lip, and all that; damage can be prevented or at least controlled if you know about the available insecticides and their proper application. Your compressor (Fig. 5-35), with either the spray gun, pressure washer, or sandblasting equipment attached, can help matters if properly operated.

When using the standard attachments, keep in mind that the spray gun should be used only to apply liquid insecticides, fungicides, crabgrass and weed killers. Large amounts of liquid can be sprayed from the material tank. A sandblasting gun, especially if it is equipped with a hopper, can be used to apply both liquid and powder (dust) types of insecticides. When large amounts of liquid are to be dispensed by the sandblasting gun from a hopper in a short period of time, it is rec-

Fig. 5-35: (A) A compact diaphragm compressor makes a fine portable garden sprayer; (B) a larger piston compressor is also portable, but with a 25 foot hose, it seldom has to be moved to spray in an average yard.

When using a canister, either feed may be used, but the pressure-feed will provide a much steadier stream of material.

In addition to spray guns and sandblasting guns, there are nozzles or wands available that work in conjunction with the material tanks. That is, the mixed insecticide is placed in the tank and when the trigger on the nozzle or wand is squeezed or depressed, either a strong stream of liquid (for spraying tree tops) or a mist (for close-up work with flowers and shrubs) is delivered (Fig. 5-36). The pressure required for nozzle work will vary from 50 to 90 psi depending on the intensity of liquid stream needed. Nozzle wands can deliver a stream about 25 feet long. The wand of a pressure washer can also be used to spray liquid insecticides on trees, shrubs, lawns, and gardens. It can also be used to deep-root feed trees and shrubs.

Because of a public concern and legal bans on hazardous chemicals, no recommendations are given as to what insecticides, fungicides, and weed killers to use with your spraying equipment. You can save a lot of trouble by checking with your local garden supply center or local agricultural extension service for the chemicals that can be used in your area. But regardless of your choice of chemicals, always read the instructions on the container before beginning to spray, and reread them every time you use the material. Go over every-

ommended that the gun be operated in the siphon-feed mode. Also, in order to obtain a high-pressure liquid stream from a hopper, it is a good idea to operate the sandblast gun in the siphon-feed mode.

Fig. 5-36: On a wand sprayer, the nozzle adjusts to deliver strong stream for treetop work (left); or a mist for close-up work with flowers and shrubs (right).

thing, taking particular note of the mixing instructions and any precautions. Then mix according to the manufacturer's directions. Do *not* use more than the recommended dosage—you could be polluting the soil. On the other hand, if a lower concentration is used, the material may not be effective and additional spraying may be necessary.

Other pointers to keep in mind on the use of dust and liquid insecticides with compressor equipment include:

1. Always mix soluble powders and granules in a pail, not in the material tank or canister to avoid clogging of spray nozzles. When thoroughly mixed, pour the mixture into these containers through cheesecloth, then stir in the recommended amount of additional water.

2. Do not overdust. A light film on both sides of the leaves is sufficient. Apply dusts on an average of every 7 to 10 days and after rain.

3. Spray or dust only when the air is calm. Drifting insecticide can be troublesome to other plants, animals, or humans. Early morning and late afternoon are often the best times because the air is usually the calmest.

4. Spray until the leaves *start* to drip. A heavy runoff indicates overspray. A few drops of liquid dishwashing detergent added to a tank of spray mixture will help the mist adhere to leaf surfaces. Also, to prevent dilution of spray solutions, spray only when foliage is dry.

5. Do not smoke or eat while spraying or dusting. Traces of spray could be introduced into your mouth by way of cigarettes or food.

6. *Thoroughly* clean all spraying equipment immediately after using it.

7. Wash all exposed parts of the body immediately. If clothes or gloves are wet with spray material, change them immediately and wash them before wearing again.

8. Keep insecticide chemicals on a high shelf out of children's reach. Better still, keep them in a locked cabinet.

There are a few simple rules to follow when spraying. Start at the lower branches and spray up and down, in a pattern of overlapping strokes. Be sure to spray on the top too. Cover all parts thoroughly so that some insecticide drips from the leaves. There is no hard and fast rule as to the amount of spray needed for good coverage. Generally, it will take 3 to 4 gallons to cover an average tree of 10 to 12 feet in height.

Fungus diseases are more prevalent during wet periods. Treatment will give control for a week provided no rain or irrigation removes the spray. Following a rain, it is vital to spray or dust again, whether for disease or for insect control. When applying fungicides, be sure to cover both sides of the leaves and stems thoroughly for complete protection.

Spraying around your home and patio (Fig. 5-37) is a good way to eliminate biting insects. It is best to spray just before an outdoor gathering so that these little pests do not come to your party. Weeds, poison ivy, and other undesirable growths can be killed by saturating the leaves with a spray of brush/weed killer. In winter, undesirable brush may be treated with a combination of brush killer, diluted with kerosene or diesel oil according to the

Fig. 5-37: Spraying around the home and patio eliminates biting insects.

Fig. 5-38: A compressor will inflate a tractor's tire whether it is of the lawn type (left) or the farm variety (right).

manufacturer's directions. Trunks are sprayed from the soil up to about a foot or 18 inches. If some of the brush remains in spring, it can then be spot-sprayed and totally eliminated.

The compressor assists in other garden work, too. For instance, it can inflate tires of your tractor (Fig. 5-38), tiller, cart, and other garden equipment. The carburetors and gas lines that become choked with grass and dirt can easily be blown out with a blow gun. The engine can also be checked for compression leaks using the same accessory. In fact, the blow gun can be used to clean away foreign matter from the entire unit (Fig. 5-39). The blow gun can be used to remove grass and clippings from sidewalks and driveways or to remove wood chips from a chain saw (Fig. 5-40). Air powered tools such as impact wrenches, drills, and grease guns can be employed for making adjustments and repairs to lawn and garden tractors and power equipment.

Fig. 5-39: Blowing grass and dirt from a tiller's engine, tines, and other moving parts.

Fig. 5-40: Cleaning a chain saw with a blow gun simplifies this important procedure.

OTHER DO-IT-YOURSELF TOOLS

As discussed in Chapter 2, several air-driven power tools—nailers and staplers—are very popular with the building trade. The reasons for their popularity is slowly being recognized by the do-it-yourselfer. These power tools can drive common nails up to 8d (2-1/2 inches) and staples of the same length. Pneumatic bradders can drive up to 1-3/4 long brads.

One of the major uses of air-powered stapling for exterior work is in re-shingling a roof (Fig. 5-41) or applying siding. The chart here gives such necessary information as staple size and type needed to hold various shingles and sidings. The staple type refers to Fig. 5-42.

The typical air-driven stapler is loaded by drawing the pusher all the way to the rear of the magazine and tilted to latch. Then insert two or three clips of staples (Fig. 5-43). To complete the loading procedure, release the pusher and let it slowly slide forward.

Spacing Specifications[4-5]	Use This Type of Staple	Fastener Gauge of Staple	Specifications[1-2] Minimum Crown	Leg Length[3]
Asphalt-Composition, Roof Shingles & Wall Shingles				
A minimum of (4) staples per each 36" section of shingle	A	16	3/4"	3/4"
A minimum of (6) staples per each 36" section of shingle	B	16	7/16"	3/4"
Asphalt-Composition, Ridge, Hip, Caps				
A minimum of (4) staples are required for ridge cap	A or B	16	3/4"	1"
		16	7/16"	1-1/4"
Roof and Wall Wood Shingles				
A minimum of (2) staples per shingle	B	16	7/16"	1-1/4"
Wood Shakes				
A minimum of (2) staples per shingle	B	16	7/16"	1-3/4"
Tin Capping—Roof Felts				
All tin caps placed and stapled 12" O.C.	B	16	7/16"	7/8"
Tin cap roofing felts to gypsum decks	B	16	7/16"	1-5/8"
Aluminum Siding[6]				
Maximum of 32" O.C.	B	15	7/16"	3/4"+

[1] Shingles and shakes attached to roof sheathing having the underside of the sheathing exposed to visual view may be attached in these locations with staples having shorter lengths than specified so as not to penetrate the exposed side of the sheathing.

[2] All staples are to be galvanized.

[3] For reroofing or recover applications, the staple leg length shall be long enough to penetrate the opposite side of sheathing 1/8" or penetrate the sheathing 3/4". All other provisions of this table will prevail.

[4] Asphalt-Composition shingles attached with staples are driven so that the staple crown bears tightly against the shingle but does not cut the shingle surface. The crown is parallel to the long dimension of the shingle course.

[5] Wood shingles and shakes attached with staples are driven so that the staple crown is parallel to the butt-edge compressing the wood surface no more than the total thickness of the staple crown wire.

[6] Staples shall be aluminum and have a minimum penetration of 3/4" into the wood supporting member. One leg of the staple shall be driven through the pre-punched hole in the sealing rib with the crown perpendicular to the width of the siding.

Fig. 5-41: A 2 hp twin-cylinder compressor, delivering 5.0 scfm at 40 psi, packs enough power to make light work of any re-shingling job.

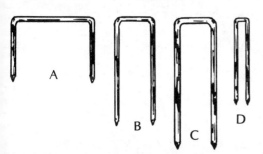

Fig. 5-42: Staples needed for various shingles and siding.

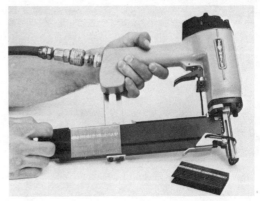

Fig. 5-43: Loading an air-driven stapler.

Normal operation of an air stapler can be usually accomplished in one of two ways:

1. Squeeze the trigger and keep it in the squeezed position, then fully depress the work contacting element (safety) against the workpiece. The tool will drive one fastener each time the work contacting element (safety) is depressed against the workpiece and released. This method is called "touch trip."

2. Depress the work contacting element (safety) against the workpiece and keep it in the depressed position, then squeeze the trigger. The tool will drive one fastener each time the trigger is squeezed and released. This is the single-cycle method. If the tool drives more than one fastener (double-cycle), you are applying too much

force in holding down the tool against the workpiece.

The nailer and bradder load and operate in much the same manner except that they drive nails and brads respectively (Fig. 5-44).

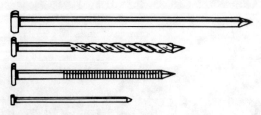

Fig. 5-44: Typical nails and brads driven by air-powered nailers and bradders.

In many cases, it may be wise to rent air-powered tools for a day or weekend to do a rarely-performed task. For instance, you hardly need to buy a stapler to shingle a small storage barn, and yet there is no point in having aching muscles that are bound to occur when nailing shingles in place by hand. Most towns have convenient tool rental services and reasonable prices. Often when you rent an air-powered tool, you can really learn of its versatility and efficiency. After renting one a few times, you may discover the advantages of ownership.

Chapter 6

INTERIOR USES FOR AN AIR COMPRESSOR

The air compressor is a very versatile machine and it does not discriminate; it helps the woman of the house as well as the man to make difficult or unpleasant tasks easier. It does this work both inside and outside the house. In the last chapter, we saw how the air compressor made various outside jobs easier; in this chapter we will look at several uses of an air compressor that make life easier by speeding up around-the-house chores so the residents have more time to enjoy their house.

Very few people like to houseclean. However, with the help of a compressor, this task can be made easier and simpler. Those difficult nooks where dust and dirt seem to grow—behind the radiators in back of the couch and TV, behind the refrigerator, and so on—can be cleaned with one blast from the blow gun which will carry the debris to where it can easily be swept or vacuumed up.

When cleaning a toaster, electric panel box, typewriter (Fig. 6-1), electric shaver, or any other appliances which cannot be immersed in water or cleaned otherwise, an air compressor will finish the job in seconds. Sandblasting, which is used primarily to remove rust (see Chapter 5), will do a fine job of polishing brass and silver if a very fine sand grit (about 90) or glass beads are used. The latter can also be used to fine polish wood. Hand polishing is usually very time-consuming, but with sandblast equipment and a compressor, it becomes a quick job.

Roof rafters in a storage building or an attic often become infested with various insects. Control and elimination of these, by using proper insecticides applied with a

Fig. 6-1: Cleaning a typewriter with a blow gun.

spray gun (Fig. 6-2), will help avoid trouble. Always wear a respirator when spraying insecticides in a confined area.

Fig. 6-2: Applying insecticides to attic or storage area rafters with a spray gun.

THE PLUMBER'S FRIEND—COMPRESSED AIR

The air compressor can help the do-it-yourself plumber with many jobs. For instance, a sink drain clogged between the fixture and the vent can usually be cleared with compressed air. To do this, remove the sink stopper and tightly wrap the blow gun with a cloth. Insert the cloth and gun into the drain, making sure the cloth plug will prevent the compressed air from escaping (Fig. 6-3). (If there is any overflow in the sink, be sure it is plugged with a cloth, too.) Trigger the blow gun and the blockage will go away with a swish.

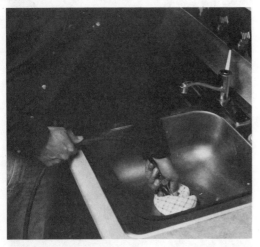

Fig. 6-3: Using a blow gun to unclog a drain line.

If this fails, it may be necessary to remove the cleanout plug at the bottom of the drain, insert the cloth plug and blow gun, and repeat the procedure. Be sure to place a pan under the trap to catch the drain water.

Compressed air can also be used to clean out water pipes, especially when closing down a summer house for the season. Turn off the water supply, open all drain cocks and faucets; and then, at the highest faucet in the system, insert a cloth plug and the blow gun. When the gun is triggered, the air pressure will force all the water out of the system, thus preventing any chance of water freezing in the lines. The same basic technique can be used to open a clogged pipe and drain or clean out a hot-water tank.

When turning a water system on, some of the older systems still require a source of compressed air to pressurize or re-prime the well tank. Pressurized-water systems are also quite common in trailers and campers.

There are times when some well water tanks become waterlogged. When this occurs, the compressor can be used to charge or re-pressurize the tank. If you have a hot-water heating system that contains an expansion tank, remember it can be re-pressurized by using an air compressor.

The blow gun comes in handy when cleaning air conditioner coils (Fig. 6-4), heating/cooling ducts, refrigerator coils, and blowing dust and dirt from the furnace burner. If you have a centralized vacuum cleaner system, there is nothing better to clean or clear blocked ducts than a few blasts of compressed air. It is also good to clear blockages in the vacuum cleaner's hose and nozzle.

Caulking Bathtubs and Showers. At one time or another, most conventionally-installed bathtubs and showers will show signs of cracking away from the adjoining tile or wallboard. This happens when the grout, used by builders to seal bathroom

Fig. 6-4: A blow gun does a great job getting rid of dust from an air conditioner's coil.

joints, begins to lose its flexibility and crumble with age. To refill the cracks with grout would only lead to the same situation within a short time. The problem can be solved by an elastic bathtub sealer which can give with the slight shifting of bathtubs, sinks, and showers that occurs as the house settles. One of the exterior caulking compounds described in Chapter 5 will serve the purpose well; however, the most popular types of compound for interior use are those which contain polyvinyl acetate or silicone.

Before sealing any joint with caulk, it must be free of debris. Remove the old grout by poking it with some sort of pointed instrument, such as an old pocket knife, a broad-edged putty knife, or a similar tool. Follow this by removing all loose particles with a wire brush. A small paint brush will allow you to make the cavity free of residue. To remove any tub soil and dirt, wash the joint with alcohol.

After the surface is completely dry and you have prepared the air-driven caulk gun as described on page 83, push the tip of the cartridge spout along the joint, applying a smooth and steady pressure on the trigger to form a solid bead. Hold the caulk gun at about 45 degrees and parallel to the joint, not at a right angle (Fig. 6-5). Any excess or uneven caulk can be smoothed with a moistened spatula or your fingertip.

If by accident some caulk should fall on the tile or a fixture, wipe it up immediately with a damp cloth or sponge. Once dry, the caulk would have to be cut or chipped away which may mar the surface. Some sealants are water-soluble before hardening; others will dissolve in mineral spirits or turpentine. Check the cartridge label for the manufacturer's instructions.

Polyvinyl acetate or silicone sealers begin to cure within several minutes and are hard enough to permit use of the sink or bathtub within the hour. After 24 hours, most sealers and caulking compounds are completely hard and totally waterproof.

Fig. 6-5: Sealing the crack around a bathtub with an air-driven caulk gun.

DO-IT-YOURSELF PROJECTS

An air compressor is a great help when it comes to working on such do-it-yourself projects as finishing off an attic or basement, paneling a room, or laying a vinyl or wood tile floor. The nailer described in Chapter 5, will save plenty of aching muscles when it comes to installing studs (Fig. 6-6) and wood framing members. The bradder can be used to drive brads which will hold finished plywood or hardboard walls.

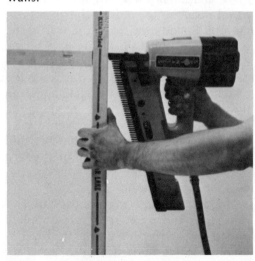

Fig. 6-6: Installing studs by employing an air-powered nailer.

The air-driven caulk gun can even be used to install wall paneling. Today the application of prefinished plywood and hardboard with panel adhesive is widely employed by the do-it-yourselfer. Its use largely eliminates the need for brads or nails and the resulting concealment of their heads. Generally, the adhesive comes ready to use in a cartridge with a plastic nozzle. This cartridge fits into the air-powered caulk gun, and the panel adhesive comes out of the nozzle as a heavy bead. If the wall is in good condition—smooth and true—the adhesive can be applied directly to the back of the panel all around the edges in intermittent beads, about 3 inches long and spaced about 3 inches apart. Keep the adhesive at least 1/4 inch from the edges of the panel and be sure that it is continuous at the corners and around the openings for electrical outlets and switches. Additional adhesive should be applied to the back of the panel (Fig. 6-7) in horizontal lines of intermittent beads spaced approximately 16 inches apart. Once the adhesive is applied, the panel may be pressed against the wall and moved as much as is required for satisfactory adjustment. To make this easier, drive three or four small finishing nails or brads about half their length through the panel near the top edge. The panel can then be pulled away from the wall at the bottom with the nails acting as a hinge. After any adjustment has been made, a paddle block should be used to keep the panel pressed back on the wall, and then the nails or brads are driven in along the edge, which can later be covered by a molding. A rubber mallet or a hammer and padded block should be used on the face of the panel to assure good adhesion between panel and wall.

This adhesive also may be used on furring strips and open studs. It is applied directly to each furring strip or stud in continuous or intermittent beads. Panels are then applied by the same method as just described above. But never apply adhe-

Fig. 6-7: Applying adhesive to the back of a plywood panel.

sives on plaster walls that have flaking paint or wallpaper that is not tightly glued. If the plaster seems hard and firm and does not crumble when you drive a nail into it, it is probably safe for adhesives. The flaking paint or loose wallpaper problem can usually be solved as detailed later in this chapter.

Another wall covering that is popular is simulated brick and stonework. For interior use, simulated plastic bricks and stones are inexpensive and easy to install using the caulk gun. In most cases they look like the real thing. But in addition to their decorative value, these textured wall surfaces can be installed without having to add bracing to the floor or a step to the foundation, which would otherwise be necessary with ordinary brick or stone because of the weight.

Imitation bricks and stones are made of various plastic materials; styrene, urethane, and rigid vinyl are the most common ones. Some are fire resistant and may be used as fasciae for fireplaces. All false bricks and stones are highly durable and come in a wide variety of colors and styles (Fig. 6-8).

Most of the better quality simulated bricks and stones are set individually, fol-

Fig. 6-8: Plastic brick and stone effects.

lowing the manufacturer's instructions. The mastic, which comes in a cartridge container, is indistinguishable from real Portland cement mortar. This mastic is usually available in white, gray, tan, or black, and it keeps the lightweight bricks and stones from slumping. The material is applied to the back of the simulated brick with the air-driven caulk gun, and once it has set, the gun can be used to fill the joints with the special cartridge mortar (Fig. 6-9). After it has set, use a dowel to clean the mortar joints.

Fig. 6-9: Applying mastic between stones with an air-powered caulk gun.

As shown in Fig. 6-10, the caulk gun can also be employed to apply adhesive to vinyl and wood floor tile. The adhesive is applied around the perimeter of a piece and then an "X" is made in the center with the caulk gun. The same technique can be used when applying ceiling tiles. Thus, the air compressor's caulking accessory can help to install the floor, walls, and ceiling of any do-it-yourself project.

Fig. 6-10: Installing floor tile.

INTERIOR PAINTING

Exterior spray painting and interior spray painting are similar in some ways but different in others. Because of the differences, the two types of painting are treated under two separate headings. Some repetition is, of course, unavoidable, but it will be kept to a minimum.

Selection of Interior Paints

Interior paints can be roughly categorized into one of three broad families: (1) flat paints that dry with no gloss and are most frequently used on walls and ceilings; (2) gloss finishes (usually used in kitchens, bathrooms, and on woodwork generally) that are available in various lusters from a low satin finish to a very high gloss; and (3) the primers, sealers, and undercoats that are used as bases. All of these paints are available in different grades or qualities, and many in both solvent-thinned (mineral spirits, turpentine, or benzine) or latex (water-thinned) forms.

Latex interior paints are generally used for areas where there is little need for periodic washing and scrubbing; for example, living rooms, dining rooms, bedrooms, and closets. Interior flat latex paints are used for interior walls and ceilings since they cover well, are easy to apply, dry quickly, are almost odorless, and can be easily and rapidly removed from applicators. Latex paints may be applied directly over semi-gloss and gloss enamel if the surface is first roughened with sandpaper or liquid sandpaper. If latex is used, follow the instructions on the container label carefully.

Flat alkyd paints are often preferred for wood, wallboard, and metal surfaces since they are more resistant to damage. In addition, they can be applied in thicker films to produce a more uniform appearance. They wash better than interior latex paints and are nearly odorless.

Enamels, including latex enamels, are usually preferred for kitchen, bathroom, laundry room, and similar work areas because they withstand intensive cleaning and wear. They form especially hard films, ranging from a flat to a full gloss finish. Fast-drying polyurethane enamels and clear varnishes provide excellent hard yet flexible finishes for wooden floors. Other enamels and clear finishes can also be used; but unless specifically recommended for floors, they may be too soft and slow-drying, or too hard and brittle. Polyurethane and epoxy enamels are also excellent for concrete floors. For a smooth finish, rough concrete should be properly primed with an alkali-resistant primer to fill the pores. When these enamels are used, adequate ventilation is essential for protection from flammable vapors.

Alternative finishing materials and their possible uses are given in the table here.

	Flat enamel	Semi-gloss enamel	Gloss enamel	Interior varnish	Shellac-lacquer	Wax (liquid or paste)	Wax (emulsion)	Stain	Wood sealer	Floor varnish	Floor paint or enamel	Aluminum paint	Sealer or undercoater	Metal primer	Latex (wall) flat	Latex gloss on sealer
Masonry																
Asphalt tile							x									
Concrete floors						x.	x.	x			x				x	
Kitchen and bathroom walls		x.	x.										x			x.
Linoleum							x									
New masonry	x.	x.											x		x	x.
Old masonry	x	x										x	x		x	x.
Plaster walls and ceiling	x.	x.											x		x	x.
Vinyl and rubber tile floors						x	x									
Wallboard	x.	x.											x		x	x.
Metal																
Aluminum paint	x.	x.										x		x	x.	x.
Heating ducts	x.	x.										x		x	x.	x.
Radiators and heating pipes	x.	x.										x		x	x.	x.
Steel cabinets	x.	x.												x		x.
Steel windows	x.	x.										x		x	x.	x.
Wood																
Floors					x	x	x.	x.	x	x.	x.					
Paneling	x.	x.		x	x	x		x	x						x.	x.
Stair risers	x.	x.		x	x			x	x							x.
Stair treads				x				x	x	x	x					
Trim	x.	x.		x	x	x		x					x		x.	x.
Window sills				x												

Note: The dot in x. indicates that a primer sealer, or fill coat, may be necessary before the finishing coat (unless the surface has been previously finished).

Estimating Paint Quantity

When determining the amount of paint needed, measure the square feet of wall area to be covered, then take these measurements to your paint dealer. He should have a chart that shows the amount of paint required for the area.

To compute the square feet of the wall area, measure the distance around the room. Then multiply this figure by the distance from the floor to the ceiling. For example, your room is 12 by 15 feet and 8 feet high. Add the dimensions, 12 + 12 + 15 + 15 = 54 feet, to find the distance around the room. Multiply this by the height of the wall to arrive at the total wall area: 54 x 8 = 432 square feet.

There are windows and sometimes doors that do not require paint, so their areas are subtracted from the total wall area. For example, in your room there are two windows, each 5 feet by 3 feet, and one door, 7 feet by 4 feet. Multiply height by width to get the square feet in each:

5 x 3 x 2 = 30 square feet of window space.
7 x 4 x 1 = 28 square feet of door space.

Total the areas of the non-paintable surfaces, 28 + 30 = 58 square feet, then deduct the sum from the total room area, 432 square feet - 58 square feet = 374 square feet of wall area to be painted. Remember, if the door is to be painted the same color as the walls, do not deduct the door area.

Besides the paint, spray gun, and compressor, you might include: extra pans or cans for stirring the paint, old cloths to wipe up spills and drops of paint, masking tape and newspaper or polyurethane sheeting to cover window glass and other surfaces to prevent smearing them with paint, a cardboard for shielding work from overspray, a stepladder or sturdy table to stand on for reaching high areas, newspapers or drop cloths to cover the floor, and cleaning materials for the gun.

Surface Preparation

Similar to exterior painting procedures, preparation for interior work depends on the surface and its condition. For example, the amount of wall preparation needed is decided by whether you are working with new walls or old ones that have had a previous treatment. If it is an old surface, the preparation needed will depend on whether the surface is plaster, wallboard, or wood, and on the type of finish used previously.

Walls Previously Papered. You can spray paint over papered walls if the paper is well bonded to the wall and contains no ink that will smear or stain. Test the paper by spraying a small section and allowing it to dry.

If wallpaper is torn, is loose from the wall in spots, or has colors that will smear, take it off and clean the walls before painting. To remove the paper, soak it thoroughly with warm water using the spray gun (Fig. 6-11). After the paper is softened, proceed to pull or tear it from the wall. If several layers of paper are on the wall, it might be necessary to take it off layer by layer. Stubborn areas are removed with a blunt-edged tool, such as a spatula or broad-edged putty knife. Be careful not to damage the wall.

Walls Previously Painted. If the previous finish is in good condition, check to see that it is free from dust, grease, or other foreign matter. A primer or seal coat is usually not needed when the painted surface is well preserved. If the old paint has a glossy finish, you may need to sandpaper the surface lightly so that the new paint will adhere.

Fig. 6-11: Removing wallpaper with the help of a spray gun.

Patching Cracks. Patch all cracks (Fig. 6-12) and nail holes in plaster walls before you apply any paint. You can fill small hairline cracks with a spackling compound or crack filler, using a putty knife to put the filler in the cracks. For larger cracks and breaks, cut out the holes and remove the loose plaster. Then cut an inverted "V", with the smallest part of the opening at the outside of the plaster surface and the largest part near the wall lath. Fill the crack with patching plaster. Dampen the edges to prevent cracking while drying. Finally, sand the patches smooth after they are dry. For most types of paint, especially the alkyd type, plaster patches must have a primer or seal coat before the final coat of paint is applied.

New Walls. Certain types of paint require the use of a primer or seal coat before you can apply them. This is especially true on plaster and wallboard because certain spots are more porous than others and will not absorb the same amount of paint. Painting without a primer will give a spotty final finish, or may produce areas that are glossy while others have a flat finish. A primer or seal coat will seal the pores of the wall surface and give an even absorption of paint.

Select a primer that best suits your needs for the type of paint you have chosen. If the primer you are using comes in white only, you may wish to add a little of your finishing paint to give a better base for the final coat.

Woodwork. Woodwork requires about the same preparation for painting as walls do. New woodwork may or may not require a primer depending on the type of paint to be applied. On previously painted woodwork, if the paint is chipped, peeling, or in bad condition, completely remove it. You can do this with a paint remover, or by sanding with an air-driven sander. If the paint is in good condition and still has a glossy look, sand off the gloss so that a new coat of paint can adhere to the surface.

To produce a smooth finish, fill the nail holes and cracks with a commercial crack filler. After the filler is dry, sand it smooth before applying the final coat of paint.

Unpainted or new woodwork to be finished with enamel or oil-base paint should be primed with an enamel undercoat to seal the wood and provide a better surface. If the unpainted wood is not primed, the enamel coat may be uneven. Unpainted wood to be finished with topcoat latex should first be undercoated. Water-thinned paint could raise the grain of bare wood and leave a rough surface.

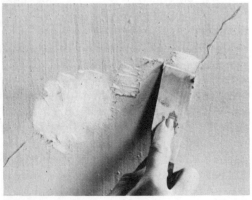

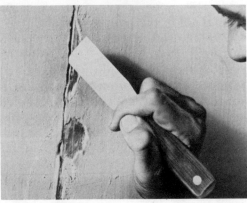

Fig. 6-12: (Top) Fine cracks in the walls and holes in the wood trim can be filled with spackling compound. Mix with water to a firm paste, and press the paste into the cracks with a spatula or a putty knife. When dry, the patched areas can be sandpapered smooth. (Bottom) Use your finger to force material into tiny cracks or corners. Large cracks will need patching plaster. On cracks 1/16 inch wide or wider, undercut to an inverted "V" shape for anchorage. Wet the edges of the old plaster so that the new plaster will bond. Mix patching plaster according to directions and fill cracks. Remove the excess and smooth the surface with a putty knife.

Application of Interior Paint

Before you spray a drop of paint, there are certain preparations you should make to do a good job with a minimum of effort, errors, and spattering. The precautions may seem obvious, but they are often overlooked. For instance, read the instructions on the label of the paint container. Then move as much furniture as possible out of room. Cover the pieces you cannot move with the drop cloths. Remove switch plates, and cover plugs and other electrical circuitry with paper. If possible, lower and cover ceiling fixtures to achieve uninterrupted ceiling coverage with the spray gun. Mask all the windows (Fig. 6-13), and block out spaces around doors to keep airborne pigment from circulating to other parts of the house. Use newspapers or drop cloths to cover the floor. Adequate ventilation is a necessity whenever you are using a paint sprayer indoors. In addition to opening windows, the forced ventilation from a small household fan directed at an open window or door will keep the room reasonably free of any paint fumes. *Caution:* When spraying indoors, *always* wear a respirator even though the windows are open.

When painting rooms, do the ceilings first, walls second, then woodwork next (doors, windows, and other trim). The place floors occupy in the sequence depends upon what is being done to them. If the floors are simply being painted, they are done last, but if they are to be completely refinished, including sanding or scraping, do them first, then cover them with paper or drop cloths while painting the room.

The basic application techniques of spray painting, described earlier in Chapter 4, hold for most interior jobs.

Ceiling and Walls. When the walls and ceilings are completely free from dust and grease, you are ready to begin. Spray the ceiling first using a 45 degree angle nozzle so that the gun and canister can be kept level (Fig. 6-14). This way, you will use all the paint and prevent excessive dusting. Consider the use of a paint tank when painting ceilings or other places where orientation of the gun is important.

Spray a strip of paint around the entire perimeter of the ceiling, using a piece of cardboard or metal as a movable mask along the junction of the ceiling and walls. Then fill in the remaining ceiling area using the large panel technique described on page 69. You will find it easier to paint the ceiling if you place a 1-1/2 inch plank at the proper height, supported securely on the treads of two solidly-grounded, completely-opened stepladders. This eliminates climbing up and down again and again. Remember that oil-base paints tend to show lap marks more than water or latex-base, so always work towards the

Fig. 6-13: How to mask a window.

Fig. 6-14: Spraying a ceiling.

wet edge when using them.

Use the same basic procedure for spraying the sidewalls that you used for the ceiling. Cover each wall separately (Fig. 6-15); use the movable mask to spray one side of the corner and the ceiling edges, then fill in with overlapping horizontal bands, as described in Chapter 4. Use the mask once more when painting near the baseboard. After the paint is dry, the baseboards are usually done in a contrasting color.

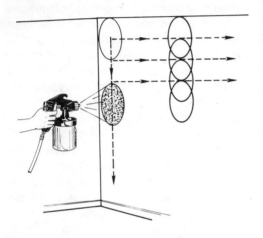

Fig. 6-15: Basic pattern of wall coverage.

Trim and Baseboards. Spray painting is a quick way to finish a window. After protecting the glass with masking tape and newspaper or plastic sheeting, adjust it so you can spray the lower part of the upper sash. Then raise the upper sash almost to the top to finish painting it. The lower sash comes next. With the window open slightly at the top and bottom, it can be finished easily. Spray the recessed part of the window frame, then the frame itself, and finally the window sill.

When spray painting a door, do the frame first, then spray the top, back, and front edges of the door itself. If the door is paneled, cover the panels and panel molding first, starting at the top by using cardboard as a movable mask. Paint the rest of the door last, again beginning at the top. Be sure to mask handles, lock-sets, and hinges. Avoid any paint buildup on or near door edges. These can cause doors to stick.

The baseboards are painted last. A cardboard or metal guard (Fig. 6-16) held flush against the bottom edge of the baseboard will protect the floor while a moveable cardboard mask should be used along the junction of the wood and wall. Do not let paper or drop cloth touch the baseboard while the paint is wet.

Fig. 6-16: Using a shield to protect the rug when spraying baseboard.

Wall or ceiling-hung cabinets can be spray painted or varnished on the spot (Fig. 6-17). Even new cabinets should be cleaned and sanded lightly before spraying. Time spent in carefully masking other surfaces adjacent to them will usually offset the trouble of removing them.

Fig. 6-17: Spraying cabinets.

Other Interior Spray Paint Jobs

Basement walls can be sprayed with any masonry, oil, or latex-base paint. If previously painted, select a paint with a similar base. If unpainted, check the label for information on priming or other surface preparation. While spraying, be sure there is adequate ventilation, or wear a mask.

Furnaces and cooling units, water heaters, stationary laundry trays, and many pieces of permanent or semi-permanent equipment in the home can be kept in beautiful condition through the use of your paint sprayer and a few minutes time. When spraying the metal housings of furnaces, central heating-cooling units, or water heaters, *be sure units are turned off* (this includes the pilot) and operating parts such as burners, blowers, and controls have been masked carefully. The surfaces to be painted should be cleaned and then sanded lightly. Depending on its condition and age, one or two coats will be needed to cover a unit.

Radiators (Fig. 6-18) and other parts of your heating system can easily be spray painted. If the radiators were previously painted, and the paint is cracked or peeling, it will be necessary to remove all of it with paint remover or a wire brush before repainting. After using paint remover, wipe off the surfaces with turpentine or paint thinner before you apply new paint. If the paint on the radiator is in good condition, all that is required is a good cleaning. Then, spray it with an oil-base paint rather than an aluminum or bronze paint. A metallic paint reduces heat output by about 10 percent. If your radiators are now covered with metallic paint, an oil-base can be applied without scraping off the old metallic paint; it is the final coat that makes the difference, not what is underneath. The oil-base paint may be any color, since the output variation from color to color is slight. Hence it is feasible to paint radiators any color desired to match your room decor. Often, radiators can be temporarily disconnected during the summer months and painted outdoors, where little or no masking is required.

Fig. 6-18: Spray painting a radiator.

Appliances, such as refrigerators or washers and dryers, can be easily refinished by the owner of a paint gun and an air compressor. Be sure to clean and dry an appliance thoroughly before spraying, and mask all hardware such as hinges, handles, and the rubber gasket around the door (Fig. 6-19).

How to Spray Furniture

Expert spraying of furniture requires a system which gets the job done with the least amount of effort and without overspraying parts already coated. Most work is best done with a small spray pattern, 3 to 5 inches wide. The small pattern is dense to the edges, permitting close control of overspray.

Fig. 6-19: When spraying a refrigerator, be sure to mask all parts that are not to be painted.

There are three common systems used for table and chair spraying. Most beginners like the four-square method, in which the operator stands opposite each of the four sides in turn and sprays everything facing him. A faster method is the diagonal system, which is done from two opposite-corner work positions. If the work has round legs, the usual system is to spray the inside of all the legs first and then spray each leg completely.

Figure 6-20 shows a typical motion study of the spraying of a small table. The work is supported on a pedestal turntable. A vertical fan pattern is used throughout. The job is started by spraying all legs on the inner edges, one surface at a time, followed by the four-square method on outer surfaces and edges. The top is sprayed last. If you are using lacquer, you may reverse the spraying direction to prevent overspraying the edges of the top, since they are already coated and nearly dry. One-directional spraying of the top is practical if you are using a slow-drying finish, since overspray on a wet coat absorbs readily. A faster system of spraying legs on two sides at a time is to use a horizontal fan for good coverage. In this case, spray the legs inside and out, and then change to the vertical fan required for the rest of the job.

In casework and cabinet jobs, the inside is always sprayed first. Then the outside is worked four-square, doing the right end first, followed by the front, left end, and top. On some jobs it is practical to leave the drawers in place, although the usual practice is to remove them to be sprayed separately.

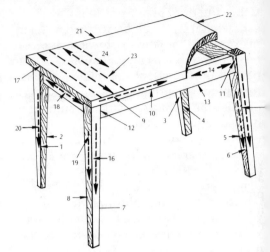

Fig. 6-20: Step-by-step details for spraying a table. Follow the numbers and arrows.

Chapter 7

THE AIR COMPRESSOR AND YOUR CAR

In these days of high costs, the do-it-yourself trend has entered the field of automotive care and repair. Your air compressor will be standing by to help solve many problems and save a great deal of time and money.

AIR-POWERED TOOLS—THE TOOLS OF PROFESSIONAL AUTO MECHANICS

The automotive industry was one of the first industries to see the advantages of air-powered tools. Today they are known as "the tools of the professional auto mechanic." Auto shop tools are the most readily available of all the types of air-powered equipment; practically every automotive parts dealer has some in stock; Fig. 7-1 shows a collection of the most popular ones. A typical air-line arrangement for the compressor in an auto shop using air-powered tools is illustrated in Fig. 7-2.

Air-Powered Wrenches

Whether changing spark plugs, removing a cylinder head, rotating tires, replacing an oil filter, doing a muffler-tailpipe job, or anything else on your car involving threaded fasteners, you can do the job much faster and easier with air-powered wrenches. There are two basic types of wrenches: the impact and the ratchet.

Impact Wrenches. The impact wrench is a portable hand-held reversible wrench. When triggered, the output shaft, onto which the impact socket is fastened, spins freely at anywhere from 2,000 to 14,000 rpm's, depending on the wrench's make and model. When the impact wrench meets resistance, a small spring-loaded hammer, situated near the business end of the tool, strikes an anvil attached to the drive shaft onto which the socket is mounted. Thus each impact moves the socket around a little until torque equilibrium is reached, the fastener breaks, or you release the trigger.

When using an air impact wrench, it is important that only impact sockets and adapters be used with impact wrenches (Fig. 7-3). Other types of sockets and adapters, if used, might shatter and fly off, endangering the safety of the operator and others in the immediate vicinity. Therefore, make certain that when you obtain sockets and adapters you specify they are to be of the impact-duty type, labeled "For Use With Impact Wrenches." A dull black-oxide finish also sets the air-powered impact sockets apart from chrome-color hand-operated sockets. Impact sockets are sold in both U.S. Standard and metric sizes.

To attach socket chucks and adapters to air-impact wrenches, merely push them onto the output shaft as far as they will go. True, power wrench output shafts come in two common variations (Fig. 7-4): the detent ball and the retaining ring. For the do-it-yourselfer, either method of attaching sockets and adapters to the output shaft works fine. Always use the simplest possible tool-to-socket hookup. Every extra connection absorbs energy and reduces power.

An adjustable air regulator is part of most pneumatic impact wrenches (Fig. 7-5). Controlling the amount of air lets you adjust the tool's speed and torque. It also allows it to be used at pressures above the usual 90 to 125 psi range without excessive tool wear. Usually numbered, air regula-

Fig. 7-1: Common air-powered tools used by the do-it-yourself auto mechanic: (A) 3/8 inch angle head impact wrench; (B) 3/8 inch ratchet wrench; (C) medium-duty short air chisel; (D) 3/8 inch palm-grip impact wrench; (E) disk sander; (F) finishing sander; (G) orbital sander; (H) 1/2 inch impact wrench with 2 inch extended anvil; (I) 1/2 inch impact wrench; (J) die grinder; and (K) 3/8 inch drill.

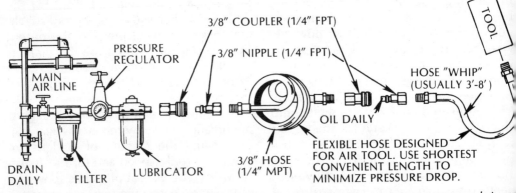

Fig. 7-2: Typical auto shop air system arrangement for tools. Details on piping compressed air are given in Chapter 8.

Fig. 7-3: The impact socket (left) is tougher than the normal socket (right). The impact type has thicker walls and a stronger six-point design.

Fig. 7-4: Two types of output shafts: detent ball (left); retaining or hog ring (right).

Fig. 7-5: The adjustable air regulator on a typical impact wrench.

tors are not so accurate that you can trust them for final torquing of fasteners. When important, final torque should be done by hand with a torque wrench. Of course, your air tank will have a regulator too. Set it at 90 psi.

Air impact wrenches work equally well for tightening and loosening. Direction of the rotation is usually controlled by a switch or two-way trigger (Fig. 7-6). Remember, do not change the direction of rotation while the trigger switch is ON.

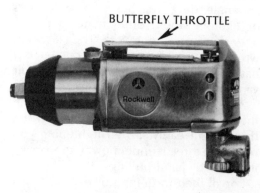

Fig. 7-6: A butterfly throttle or switch acts as a forward and reverse lever for this palm-grip air impact wrench, permitting either a forward right-hand rotation or a reverse left-hand rotation.

To remove fasteners, set the switch for left-hand rotation. Place the socket over the nut or fastener head. Exert forward pressure on the wrench as you depress the trigger switch. As soon as the nut or fastener becomes loosened, relax the forward pressure on the wrench to let it spin the nut or fastener free.

To install fasteners, set the switch for right-hand rotation. By hand, start the nut on the stud or the bolt on the threads; this will help to avoid crossthreading which would ruin the fastener. Place the socket over the nut or fastener head. Depress the trigger switch to drive the nut or fastener until it rests on the material being fastened, and then exert forward pressure on the wrench to bring the hammer into action to snug the nut or fastener firmly (Fig. 7-7). Deep-flex sockets permit out-of-line fasteners to be reached without time-consuming socket and extension changes. When loose, the bolt drops into the socket.

Air Ratchet Wrench. This air ratchet wrench, like the hand ratchet, has a special ability to work in hard-to-reach places. It is good for handling spark plugs, oil pans, manifolds, carburetor mounting bolts, locking the belt adjustment on alternators, radiators, and other bolts that are difficult to get at and turn any other way (Fig. 7-8). Its angle drive reaches in and loosens or tightens where other hand or power wrenches just cannot work. The air socket wrench looks like an ordinary ratchet, but has a fat handgrip which contains the air vane motor and drive mechanism (Fig. 7-9).

After breaking loose a fastener by pulling the air ratchet handle, you can power the threads out easily. When tightening you run in the threads under power, then make it secure by hand-pulling.

For all their torquing power, air impact wrenches have practically no recoil. Holding one is rather easy. On the other hand, the air socket can bust your knuckles when the turning gets tough. The air socket must be tightly held.

Fig. 7-8: Loosening a nut using an air ratchet wrench.

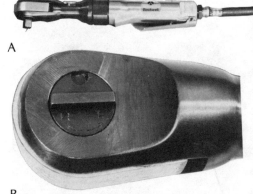

Fig. 7-9: (A) A typical air ratchet wrench. The forward and reversing lever (B) is used to change the rotation of the tool. When the lever is set in the forward position, the tool will turn in a right-hand rotation to fasten nuts and bolts. When the lever is moved to the reverse position, the tool will run in reverse or left-hand rotation to remove nuts and bolts.

Remember that there is no consistently reliable adjustment with any air socket or impact wrench. Where accurate preselected torque adjustments are required, a standard torque wrench should be used. The air regulator on air-powered wrenches can be employed to adjust torque to the approximate tightness of a known fastener; a 75 foot pound torque handles most automobile jobs. *Note:* Actual torque on a fastener is directly related to joint hardness, tool speed, condition of socket, and the time the tool is allowed to impact.

Fig. 7-7: Installing a nut to the wheel using an impact wrench.

Air Drills

Air drills are usually available in 3/8 and 1/2 inch sizes and operate in much the same manner as an electric drill. But, as can be seen in Fig. 7-10, they are smaller and lighter. This compactness makes them a great deal easier to use for drilling operations in auto work (Fig. 7-11).

To drill with an air tool into any material, the following general procedure should be kept in mind:

1. Accurately locate the position of the hole to be drilled. Mark the position distinctly with a center punch or an awl to provide a seat for the drill point and to keep it from "walking" away from the mark when you apply pressure.

2. Unless the workpiece is stationary or large, fasten it in a vise or clamp. Holding a small item in the hand may cause injury if it is suddenly seized by the bit and whirled from your grip. This is most likely to happen just before the bit breaks through the hole at the underside of the work.

3. Carefully center the drill bit in the jaws as you securely tighten the chuck. Avoid inserting the bit off-center because it will wobble and probably break when it spins. After centering, place the drill bit tip on the exact point at which you wish to drill the hole, then start the motor by pulling the trigger switch. (Never apply a spinning drill bit to the work.)

4. Except when it is desirable to drill a hole at an angle, hold the drill perpendicular to the face of the work.

5. Align the drill bit and the axis of the drill in the direction the hole is to go and apply pressure only along this line with no sidewise or bending pressure. Changing the direction of this pressure will distort the dimensions of the hole, and it could snap a small drill. To avoid stressing the drill bit, try extending your index finger along the side of the drill housing, with your middle finger on the side.

6. Use just enough steady and even pressure to keep the drill cutting. Guide the drill—*do not force it*. Too much pressure

Fig. 7-10: A 3/8 inch air drill (left) and 3/8 inch electric drill (right). The air drill weighs 2-1/2 pounds, while the electric type weighs 4-1/2 pounds.

Fig. 7-11: An air drill at work.

may cause the bit to break or the tool to overheat. Too little pressure will keep the bit from cutting and dull its edges due to the excessive friction created by sliding over the surface.

7. When drilling deep holes, especially with a twist drill, withdraw the drill several times to clear the cuttings. Keep the tool running when you pull the bit back out of a drilled hole. This will help prevent jamming.

8. Reduce the pressure on the drill just before the bit cuts through the work.

Air Sanders. There are two basic types of air sanders: disk and finishing (pad). Most sanding done in automotive work is done with a disk sander or its counterpart, the orbital sander. The latter has an orbital action rather than the circular one of the disk sander. This orbital action gives a swirl-free finish; swirls can be a problem with disk sanders. Grinding wheels and wire cup brushes are available accessories for some disk sanders.

When using a disk sander, keep the following points in mind:

1. Avoid grinding or sanding too close to the trim, bumper, or any other projection that might snag or catch the edge of the disk.

2. Do not *stop* a disk grinder while in contact with the work surface. *Start* the machine *just* before you contact the work surface.

3. Never remove any safety guards that might be part of the sander.

4. Grinding should direct sparks and dust away from the face and toward the floor.

5. If a sander is held at too sharp an angle, it will have a tendency to skip out of control (Fig. 7-12). Hold the sander at a low angle to the work surface (5 to 10 degrees).

6. There are a number of things to check concerning your backup pads. Make certain they are free of cuts or nicks at the edges and at the centerhole. Check for proper balance—excessive vibration may indicate an out-of-round backup pad. The retainer nut must have at least a three-thread contact.

7. Disks should not over-hang a backup pad. Do not use cloth-backed sanding disks. Use only paper disks and proper adhesive. Also make sure the rpm rating of the pad corresponds with the manufacturer's recommended speed for the tool.

8. Start the work with an abrasive grit just coarse enough to remove high spots and roughness. Follow with additional sandings using finer grits until the desired finish is obtained; never go from a coarse grit to a fine grit in one step. It may be difficult to remove swirl marks made by coarse abrasives. To prevent marks, use the finest grits practical for roughing operations and finish by using successively finer grits.

Fig. 7-12: A disk sander (top) and orbital (bottom) sander in action.

9. Let the sander do the work. The normal weight of the machine is sufficient for efficient sanding. Do not put additional pressure on the machine. This will only slow down the speed of the pad, reduce sanding efficiency, and put an additional burden on the motor. Start the sander off the work, then set it down on the work evenly and move it slowly back and forth in wide, overlapping areas. When finished sanding, lift it off the work before stopping the tool.

The finishing sanders, also called pad sanders (Fig. 7-13), are designed for fine finish sanding—as their name implies. It is possible to use a wider variety of abrasives with finish sanders than with any other type of power sanding, but for the most

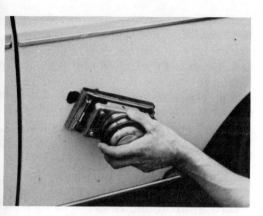

Fig. 7-13: An air finishing sander at work.

part, the best work is done with comparatively fine grit abrasive paper. Finish sanders are also especially designed for sanding hard-to-reach places and tight corners.

Begin the sanding operation with an abrasive grit just coarse enough to remove the high spots and any excessive roughness. Follow with a second sanding using a grit one or two grades finer. Continue with successively finer grits until the desired finish is obtained. Do not go from a coarse grit to a very fine grit in one step. By so doing, it will be impossible to remove swirl marks that might have been made by the coarse grit abrasive. Always use the finest grit practical for the roughing operation.

Polishing and smoothing metal, plastic, painted, and some other similar surfaces requires a lubricant such as water, water-soluble oil, or another noninflammable liquid. When selecting the abrasive for use with lubricants, be sure it has a waterproof paper or cloth backing, usually called a wet-or-dry type abrasive sheet. Sponge the liquid onto the surface leaving a very light film. Do not flush it on.

When doing any sanding or grinding operations on cars, be sure to follow these safety rules:

1. A filter face mask should be worn to minimize dust inhalation.
2. Wear safety glasses or goggles with side shields when straightening, grinding, or sanding metal or plastic.
3. Be careful of loose clothing such as unbuttoned sleeves or loose jackets.
4. Wear gloves whenever possible.

Other Air Auto Tools

One of the most important "other" auto tools is the air chisel (Fig. 7-14). Used with the accessories listed, this tool will perform the following operations:

1. *Universal joint and tie rod tool.* They help to shake loose stubborn universal joints and tie rod ends.
2. *Smoothing hammer.* A good accessory for reworking metal.
3. *Ball joint separator.* The wedge action breaks apart frozen ball joints.
4. *Panel crimper.* It forms a "step" in a panel where a damaged section has been removed. The "filler" panel will then fit flush, resulting in a strong, professional joint.
5. *Shock absorber chisel.* Quick work of the roughest jobs is made without the usual bruised knuckles and lost time. It easily cracks frozen shock absorber nuts.
6. *Tail pipe cutter.* The cutter slices through tail pipes and mufflers.
7. *Scraper.* Removing undercoating in addition to other coverings is this accessory's function.
8. *Tapered punch.* Driving frozen bolts, installing pins, punching or aligning holes, are some of many uses for this accessory.
9. *Edging tool (claw ripper).* It is utilized to slice through sheet metal leaving a smooth edge.
10. *Rubber bushing splitter.* Old bushings can be opened up for easy removal.
11. *Bushing remover.* This accessory is designed to remove all types of bushings. The blunt edge pushes but does not cut.
12. *Bushing installer.* The installer drives all types of bushings to the correct depth. A pilot prevents the tool from sliding.

In addition to special chisels, there are the so-called "standard" types which can be used for cutting rivets, nuts and

Fig. 7-14: An air chisel at work.

bolts, plus removing weld splatter and breaking spot welds.

A die grinder is a great tool for wire cord tire repairs (Fig. 7-15). It can also double as an excellent woodworking tool (see page 149).

Fig. 7-15: An air die grinder at work.

The pressure washer described in Chapter 2 provides an excellent means of cleaning car bodies, engines (Fig. 7-16), machinery, and even floors. Used with the flushing adapter, it can flush out a car's radiator. *Note:* When degreasing automobile engines or any other engines with air/water pressure, make sure carburetor throat is properly covered. Disconnect all electrical sources. Remove the battery.

The blow gun is another handy cleaning tool around an auto shop. Blowing away debris from around the engine, cleaning out gas lines and choked carburetors, and removing dead bugs from behind the radiator (Fig. 7-17) or grille are only a few of its many cleaning uses. The blow gun can also be employed to check an auto engine's compression level and the operation of the air conditioner. The sandblast gun can be used to remove rust from the body of a car and to decarbonize engine parts.

The work of greasing trucks, tractors (Fig. 7-18), cars, and other equipment that needs lubrication will be made much simpler and quicker with an air-operated grease gun. As the owner of various kinds of vehicles, you can benefit from a compressor with a tire chuck that can keep all tires properly inflated for longer wear (Fig. 7-19). And a flat tire on a vehicle at home need not be a problem. If you do not

Fig. 7-16: Cleaning a car body (left) and its engine (right) with a compressor powered pressure washer.

Fig. 7-17: Removing bugs from behind a radiator with a blow gun.

Fig. 7-18: An air grease gun will do a good job greasing all vehicles.

Fig. 7-19: An air compressor will keep your tires properly inflated.

There are some specialized air-powered auto mechanic tools that are found in commercial garages. These tools—their names usually imply their use—include a radiator tester, cylinder hoist, jack hoist (Fig. 7-20), air filter cleaner, body polisher, brake tester, transmission flusher, engine cleaner, and spark plug cleaner. They are, however, seldom found in a do-it-yourselfer's garage.

Remember that most air tool motors need daily (or as often as used) lubrication with a good grade of air motor oil. If the air line has no line oiler or lubricator, run a teaspoon of oil through the tool. The oil can be squirted into the tool air inlet (Fig. 7-21) or into the hose at the nearest connection to the air supply; then run the tool. Most air tool manufacturers recommend the use of their special oil for lubricating tools; however, when this is not available standard automatic transmission fluid may be substituted.

Fig. 7-20: An air jack hoist at work.

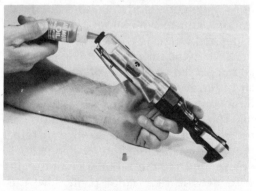

Fig. 7-21: Hand oiling of an air powered tool.

wish to fix it yourself, the air compressor will restore its inflation temporarily, until you get to a gas station or garage.

REFINISHING A CAR

Whether you want to paint a fender or the entire body of an automobile, a spray gun with a compressor is an absolute necessity for maintaining the original luster-finish quality. For small dents and minor repairs, an air brush may be used.

Repairing Minor Scratches or Chips

To repair minor scratches or chips, proceed as follows:

1. Clean the surface around the chipped or scratched area with soap and water. Apply a wax and grease remover to remove silicones, wax, and grease which may cause paint failure.

2. Wet sand the area with very fine sandpaper or a sanding block to remove the scratch and taper the surrounding paint edge. Make sure the surface is clean and dry before applying the automotive primer with an air brush. Mask off the trim and adjacent areas with masking tape and paper to avoid overspray. Apply primer according to manufacturer's instructions.

3. When the primer dries, lightly sand the surface with very fine sandpaper and water. Let it dry, then apply plastic or glazing spot putty with a spreader to cover all scratches and imperfections.

4. When the spot or glazing putty is dry, sand it smooth with extra fine wet-or-dry abrasive paper and water. Wipe the surface clean and allow it to dry before applying two or three coats of primer, according to the manufacturer's instructions. When dry, sand the primer to a final smooth surface with super fine abrasive paper and water.

5. When the surface is dry, wipe it clean and apply the color coats of paint with an air brush, according to the manufacturer's directions. Two or three days after painting, rub the area with rubbing compound to bring out the sheen. Be sure to refer to the paint manufacturer's instructions before applying the rubbing compound.

To get the right color paint, give the auto parts dealer your car manufacturer's paint code. To find the tag or plate that names the code, look in these places:

American Motors: left front door body hinge pillar, or on rear face of left front door.

Chrysler: left front wheel well housing under the hood, or next to the windshield inside the car.

Dodge: left front fender-side shield or wheel housing.

Plymouth: right or left front fender-side shields, left or right side of cowl panel, left side of radiator side panel member.

Ford: rear face of left door.

On trucks, campers, vans, and GM vehicles: check the glove compartment, or ask the parts store.

Repairing Minor Dents

Minor dents are those that do not require special metal straightening tools and equipment. To repair a minor dent, proceed as follows:

1. Clean the damaged area first by washing it with soap and water. Wipe it dry and then use wax and grease remover. Failure to completely clean a surface that is later painted could require resanding and repainting. An air-powered pressure washer will help to make this washing task easy.

2. When the dent is more than 1/4 inch deep, you should restore it as closely as possible to the original shape. You may be able to lightly hammer it out from the inside, but it may be inaccessible so you must pull it out from the outside. Drill a small hole (use about 1/16 inch bit) in the center of the dent or at the point of original impact (Fig. 7-22). Partially screw in a sheet metal screw that is slightly larger in diameter than the hole, allowing enough to protrude so the head can be gripped with vise-grip pliers. Pull gently until the dent comes out. If needed, drill additional holes and insert screws, continuing the process until the dent is pulled out.

An alternative method is to use a small dent puller or a "slide" hammer. Avoid overpulling. The final smooth contour will be obtained using a plastic filler.

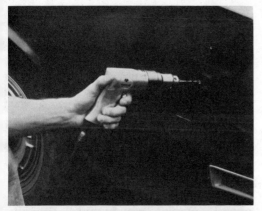

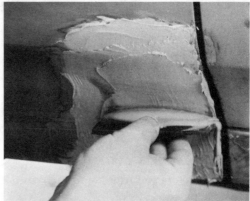

Fig. 7-22: To remove a small dent, first drill a small hole using an air drill, then employ a metal screw and pliers to pull the dent out.

Fig. 7-23: Apply the plastic filler to the damaged area (top) and after it dries completely, sand it smooth (bottom).

3. Remove all the paint and primer down to the bare metal. Clean to at least 1/2 inch around the dented areas with coarse abrasive paper on an air disk sander.

4. Mix and apply plastic filler according to the manufacturer's directions. After the filler has hardened, sand the area with medium grade abrasive paper on an orbital air sander or a sanding block (Fig. 7-23). Finish sanding the plastic filler and featheredge or taper the paint surrounding the repair area with fine abrasive paper. Figure 7-24 demonstrates the principle of featheredging.

5. Make sure the surface is clean and dry before applying the automotive primer with either air brush or spray gun. Mask off the trim and adjacent areas with masking tape and paper to avoid overspray.

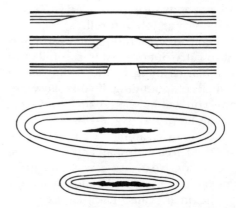

Fig. 7-24: It is very important to properly featheredge the area that is going to be repainted. These drawings show how a featheredged spot should look. The wider the featheredge around the repair, the better it will be. The objective of featheredging is to get a very gradual taper from the old paint to the bare metal.

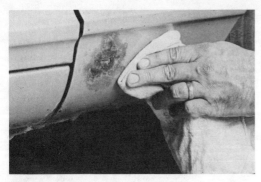 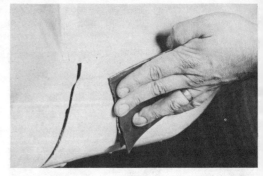

Fig. 7-25: Before and after photos of a rustout repair job.

Apply the primer according to the manufacturer's instructions.

6. For application of spot putty and repair, refer to the previously detailed procedure for minor scratch repairs beginning with Step 3.

Repairing Minor Rustouts

What is a minor rustout? Rusted areas that are seen on the surface may seem small in size but some of the adjacent metal is probably also rusted and very thin and weakened. If the rusted area is larger than 5 by 7 inches, it should probably be cut out and new metal welded or pop riveted in place. A new or used panel, fender, rocker panel, or door may be necessary. These large jobs should be taken to a professional body shop; however, you can do the refinishing job yourself.

When you repair rustouts it is very important to carry out every step carefully from the initial cleaning of the rusted area to the final compounding. If you follow these steps, you will have a repair job of which you can be proud (Fig. 7-25).

1. Begin by cleaning the surface around the rusted area with soap and water. Then wipe it dry. Next, use a wax and grease remover to dissolve silicones, grease, or wax that might cause paint failure.

2. Sand away the rust and paint to one inch around the rusted area with coarse grade abrasive paper mounted on the air disk sander. Make sure at least 1 to 1-1/2 inches of the solid bare metal is exposed around the rusted area. *Note:* The feather-edge is complete when you can run your fingers over the area without feeling any edges between bare metal, primer, and finish.

3. With a small hammer, depress or slightly bend in the edges of the solid metal around the repair area (Fig. 7-26A). This is necessary so the backup patch will be recessed to allow for plastic filler buildup. Applying enough plastic filler lets you sand it down to obtain the original contour.

4. To bridge the repair area, use a piece of self-sticking body patch which is available at most auto supply dealers. Use a pair of scissors to cut a patch large enough to cover the entire sanded area. Moisten your fingers, remove the clear film backing from both sides (Fig. 7-26B) and apply the sticky side to the surface, stretching it across the repair area. Press the patch down firmly where it contacts the metal (Fig. 7-26C). Smooth it out to remove any air bubbles or wrinkles. Moisten your fingers again with water; keeping your fingers wet prevents the patch from sticking to your hands.

5. The curing time depends on temperature and available sunlight. The table gives the approximate curing times:

Temperature	Bright Sunlight	Partly Sunny
100 degrees	1/2 - 1 hour	1-2 hours
80 degrees	1-2 hours	2-8 hours
50 degrees	4-12 hours	8-24 hours

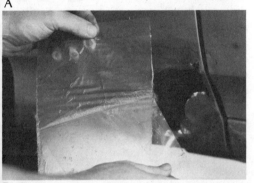
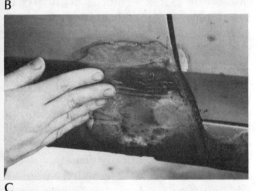

Fig. 7-26: Steps in applying a body patch.

The patch can also be cured indoors by using an ultraviolet sunlamp. Heat alone will not cure the patch; direct rays from the sunlamp must strike it. Place the sunlamp 4 to 6 inches from the patch. Do not allow the patch to over-cure, as this may cause brittleness. Usually 12 to 20 minutes under the sunlamp is sufficient to cure a patch (Fig. 7-26D).

6. After the patch has hardened, scuff sand the surface with a medium grade sandpaper disk. Then apply the plastic filler following instructions given in the previous section for repairing minor dents, beginning with Step 4.

Complete Refinishing Job

When completely refinishing a car, you have a choice of finishing over bare metal or over a factory surface. Let us examine how each is handled.

Painting Over Bare Metal. This approach is used mainly when the surface of the vehicle is brought down to bare metal by grinding, sanding, or stripping. This is the ultimate method, but is rarely used, as much custom paint today is laid over stock, re-primed, or sealed surfaces with a basic layer of paint underneath. Before painting over bare metal, the entire factory coat of paint or primer is removed. It can be sanded to the bare metal by hand, machine, or stripped with a paint stripping liquid. The sanding method speaks for itself. When using liquid stripper, the liquid is applied, usually by brush, and allowed to remain on the surface. By chemical action, the painted layer bubbles and crinkles off as the chemical action softens and separates the paint film from the surface. The use of a scraper is sometimes necessary to release paint that clings to the metal even after several applications of the stripper. If rust spots appear, they should be sanded to bare metal. Sandblasting is another excellent method for taking paint down to base metal. It can be easily done with the equipment described in Chapter 2 and the technique detailed in Chapter 5.

A good metal prepping agent should then be used to wash down the base steel and to inhibit rusting. Finally the metal should be wiped with wax and grease remover to prepare the surface for the primer. Sanding the surface to as smooth a finish as possible will help to keep the new paint faultless and scratch-free.

After preparing the metal, the first major step in painting is primecoating. Most paints are not designed to adhere to bare metal. They may stay on for a short period but will eventually become brittle—their bonding capabilities destroyed. Good adhesion is mandatory and for this purpose, primer, or primer surfacer, especially designed to offer optimum adhesion, is applied to the bare metal surface. All automotive refinishes, including enamel, acrylic enamel or acrylic lacquer, need a proper prime coat foundation. Primer also minimizes sand scratches and acts as a "filler." In cases where scratches are still prominent after priming, the use of a glazing putty is advised. Glazing putty is applied over and worked into the scratches in light but adequately-filling coats. Glazing putty is very thick and well suited for filling in deep scratches. After it has dried, the puttied area is sanded until the scratches have been feathered out, then the area is re-primed. This process is repeated until the area to be painted is smooth, flawless, and ready for overcoating. Prior to color painting, the primed surface may be wet sanded with extra fine and super fine abrasive paper to get a perfectly smooth painting surface.

The majority of repaint jobs today are done over stock factory finishes. This is the simplest and cheapest method...and flawless if the proper procedure is exercised. Sealers must be used when painting over old surfaces, factory finishes, or undercoats of a nature or solvent base differing from the new paint being applied. Sealers are basically designed for two main purposes:

1. To guard against solvent penetration of new paint into old surface paint. (This is especially important when lacquers are sprayed over enamels.)

2. To prevent adverse reactions between new and old coats such as blistering, peeling, and bad adhesion.

Penetration of the new paint solvents also tends to swell sand scratches on the old paint and can be prevented by sealing. When thinner evaporates out of the under-surface scratches or blemishes, the new coat of paint sinks in causing scratched, dull finishes. Penetration also results in lack of uniform gloss. A good two coats over the entire area to be repainted will inhibit the new coat penetration of the old undercoat. The blocking coat of sealer will allow the finish coat to retain a good, even gloss with no solvents allowed to escape into the undercoats.

The technique of overpainting is far more simplified than the bare metal approach, and also allows the painter to use the vehicle's factory color as an undercoat. The following procedure illustrates how easy it is to revamp your car by overpainting (Fig. 7-27):

1. Degreasing is the first step. Mud and oil sludge that has collected around the wheel wells, and the surface silicone laid on with waxing and buffing compounds must be eliminated. A good heavy strength wax and grease remover works well here.

2. All emblems, stickers, labels, and exterior light fixtures are removed wherever possible (Fig. 7-27A).

3. The entire vehicle is then swabbed down with a wax and grease solvent, thoroughly removing all dirt and film (Fig. 7-27B). A number of applications may be necessary to completely break down dirt and road film. The wax and grease remover is then wiped off completely and surfaces dried with cotton rags. Polyester rags and cloths are not recommended as they may be broken down by the solvent and their chemicals worked into the surface.

4. A thorough washing with mild soap will also help to completely remove the

Fig. 7-27: Steps in prepping a car for paint over a factory surface.

wax and grease solvent (Fig. 7-27C).

5. After cleaning and prepping, mask all door jamb areas. This prevents overspraying and paint blowing through cracks (Fig. 7-27D).

6. Areas to be painted are masked off, using tape and paper (Figs. 7-27E and 7-27F).

7. All the surface area to be painted is wet sanded thoroughly but carefully with extra fine grit wet sandpaper (Fig. 7-27G). The sanded surfaces are then wiped clean with a damp cotton cloth and dried with cotton material. Wiping with a tack rag removes lint and dust.

8. Two coats of clear sealer are then applied on the surfaces to be painted. Two hours after the sealer is applied, the vehicle is masked and ready for repainting.

9. Sand under fenders and prime fender wells (Fig. 7-27H).

Spray on one or two finish coats of paint. To avoid lapping, spray a full wet coat on each panel before going on to another section of the car. For example, spray the complete width of the car top in one section, if possible. Do not spray a section too heavily as this will cause runs and sags. Should this happen, sand the section down to a smooth meeting surface with surrounding area.

Let the new finish fully harden for at least two weeks before attempting to rub down the surface with rubbing compound or

Fig. 7-28: Customizing a van with an air brush.

applying a wax coating. In most cases with modern auto finishes, the final protective wax coating is all that is needed. Always check with the manufacturer's instructions before using a rubbing compound.

Customized Painting

Customized painting is in vogue today, especially in the case of vans. Most of this work is done with an air brush (Fig. 7-28). For complete details on this subject, we suggest the Badger book entitled *Air Brushing Techniques For Custom Painting,* which is available at local art supply stores or by writing to the Badger Air-Brush Company, Franklin Park, Illinois 60131.

Chapter 8

THE AIR COMPRESSOR AND YOUR HOBBIES

Using the air compressor with its many tools and accessories, which take much of the work out of the various burdensome tasks in and around the home as described in the previous chapters, you will have a great deal more time to pursue your favorite hobby or leisure activity. Here the air compressor will help you again by making your avocations easier to enjoy. If you are the sportsperson who likes to play touch football, basketball, or go bike riding, the inflator kit accessory is ideal for keeping your air-inflatables working (Fig. 8-1).

Bicycle frames or other items on wheels, such as mini-bikes and toys of almost any type and size, can be refinished with your spray gun (Fig. 8-2)—a job you would find very time-consuming if you had only a brush to apply the paint. First, mask off moving parts and bright-plated areas. Be sure surfaces and frame members are clean and dry. After each coat, be sure to allow sufficient time for complete drying before applying another.

Fig. 8-2: Refinishing a mini-bike.

FOR ARTISTS, MODELERS, CERAMISTS, AND OTHER CRAFTSPERSONS

For the craftsperson and hobbyist, the air brush working in conjunction with a compressor is a vital tool. For the hobbyist or fisherman who wants to fabricate and design his own lures, it is a must. Most of the finely detailed fishing lures on the market today are painted and decorated with the air brush (Fig. 8-3). The fine spray pat-

Fig. 8-1: For any sportsman, an inflation station such as this is ideal.

Fig. 8-3: You can use either model enamels or acrylic lacquers to paint lures. Clear urethane can be sprayed over the design for additional protection. The eyes are brush-detailed by hand.

A

B

Fig. 8-4: (A) The outer edges are traced and fogged with the air brush; (B) multiple overlapping produces this effect. Freehand swirls are added. Dots and soft edge circles may be added also.

tern of the air brush makes it the perfect tool for the tiny detail required on plugs and spoons. Quick-drying lacquers are used since they are generally impervious to water, moisture, and discoloration. The air brush is also an excellent tool for redoing used lures.

Actually air brush uses are limited only by your imagination. Artists are discovering many ways to use the air brush to achieve new artistic freedom. For instance, modern op-art designs are simple to achieve. Op-art stencils can be hand cut or art aids, such as engineering stencils or French curves, can be used to obtain dramatic effects (Fig. 8-4).

An increasingly popular art form is T-shirt air brushing. With a few hours of practice, the air brush is as easy to use as a paint brush or a pencil. The design can be applied by using a stencil (Fig. 8-5) or it can be done freehand (Fig. 8-6).

Mixing Paints

There are myriad pigments and paint types that can be used in air brushing, and since they vary in density and consistency, this guide on mixing proportions is provided. Because of the small spray opening of the air brush, the paint must be thin enough to properly atomize into a working, trouble-free spray pattern. The following is a list of some of the more widely used paints and their recommended paint-to-thinner ratios:

Fig. 8-5: Applying a stenciled design to a T-shirt with an air brush.

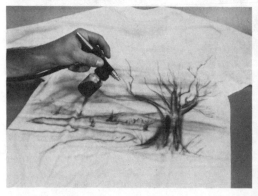

Fig. 8-6: Applying a T-shirt design by the freehand method.

Silk Screen and Printing Inks. Silk screen and similar waterproof printing inks are popular for air brushing, especially in fabric tinting and painting (T-shirts, batik, and so on). Since these inks are also very viscous, they require much thinning. Screen inks are water soluble and a one part ink to three or four parts water make good, workable spraying solutions.

Floquil and R. R. Lacquers. These flat lacquer paints are especially suited for painting metal. They are of a moderate consistency and need not be thinned as much as other mediums. One part lacquer to two parts thinner gives an excellent atomizing solution.

Airplane Dope. Used almost exclusively on flying and static models, dope is fast drying and will air brush well. Ratios of one part dope to three parts thinner or one part dope to four parts thinner are recommended. The thinner to be used here is acetone.

Model Enamels. For model finishing on boats, planes, and trains, moderately fast-drying enamels are widely used. Enamels are easy to use and ideally suited for air brush spraying. Though specific thinners are produced by the company, most of these enamels can also be thinned with inexpensive lighter fluid or turpentine. Recommended thinning ratios are one part paint to one or two parts thinner. The European-canned paints are of a thicker consistency and require a three-to-one thinning ratio.

Tube Water Colors (Gouache). Gouache colors are similar to caseins and are water soluble. They must be well-mixed until broken down to a watery consistency. One part paint to four and five parts water afford the best air brush rendering consistencies. Full and intermittent cleaning during air brushing is recommended to keep the workings of the air brush and spray orifices free of paint coagulation.

Tubed Acrylics. The newest of commercial and fine art paint mediums—tubed acrylics—may be used with care for air brush rendering. Because of the extra fast drying that is characteristic of the acrylic in stock state, it must be well-thinned and the air brush must be constantly cleaned and flushed, as acrylics clog and block up more than any other medium with the exception of the Toluene-based automotive acrylic lacquers. A one-to-five paint thinner mix is advised and best suited for water-based acrylic spraying.

Acrylic Lacquer. Most popular in the automotive custom and "specialty" paint field acrylic lacquers also have diversified application in the commercial and finer art fields. They dry instantaneously; they are the acknowledged fastest drying paint liquids, second to none in set-up speed. These lacquers are fully described in Chapter 7.

For the Modeler

The properly executed model should not only be flawlessly painted but have a refined surface as well. In the case of plastic scale models, the finished model is the sum total of all its parts. Since some deviations in the injection-molded parts are bound to occur, the parts must be properly finished and in some cases modified. When parts are oversized and will not fit into designated areas, they can be trimmed down with a hobby knife or sandpaper. When parts are shorted, gaps and spaces will appear which should be corrected. Most models made today require a touch of putty somewhere. Filling in or puttying is mandatory if you wish to achieve show-model quality. Puttying is relatively easy, and the materials involved work well in experienced hands as well as inexperienced.

Two products that work best are contour putty and glazing putty. Contour putty is designed specifically for model making (Fig. 8-7). Glazing putty is an automotive paint filler used to conceal and feather out scratches on primed automobile surfaces (see page 131). Both of these products work well, but the auto glazing putty is easier to work with since it dries faster and sands

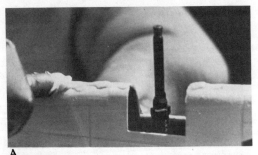

Fig. 8-7: (A) Contour putty is applied as shown, smoothed out, and (B) allowed to dry. It should be applied sparingly and built up slowly.

Fig. 8-8: Giving an auto model that metallic look.

more readily than contour putty. This material should be built-up slowly with sanding between layers until gaps and flaws have been refined. In some cases where the buildup of putty obscures body lines embossed in the plastic, the lines may have to be re-etched with a scribe, awl, or hobby knife.

When masking is required, tape is the most popular method. The paper tape, used for automotive and paint masking, is preferred because it can be easily applied to curves and circles and does not absorb paint. The tape should be burnished down to prevent loose edges where paint can creep underneath.

Metallic enamels used in model work (Fig. 8-8) are difficult to manipulate since they tend to streak and blotch when applied with even the best of brushes. The precise operation of the air brush enables you to spray a flawless, even paint job on any model, large or small. Enamels direct from the bottle are too thick for proper spraying and atomization through the narrow openings of any air brush. The stock paint should be thinned so that it will spray on evenly. Normally, a one-to-one thinner-to-paint ratio is a good mixture; however, more thinner may be needed for proper spray flow. Metallics should be sprayed in even, slightly overlapping side-to-side strokes. Shake the paint bottle frequently but not violently to keep the metallic particles suspended and insure an even coating. Allow each coat sufficient drying time before applying the next one.

No matter how detailed a model, it is not authentic looking without being weathered (Fig. 8-9). Dirt, grime, grease, and soot are all present even on the newest real life version. Most serious modelers in the war, railroad, and diorama categories weather their projects. The degree of weathering is up to the modeler. The scale replica may be slightly aged or extremely battered. Several companies produce a line of special weathering and aging colors which provide all forms of wear and decay. Though other means exist, weathering and aging effects are best achieved with an air brush. The fogging and hazing techniques of spraying make the air brush the ideal weathering tool, and the air brushes with their finely-controlled settings enable you to master the technique easily. Before attempting to weather a new model, practice on a card, paper, or an old model so that fine control is developed before applying

Fig. 8-9: A weathered finish applied by an air brush gives any model an authentic look.

weathering colors to the new project.

There is no better way to apply camouflaging to war models (Fig. 8-10) than with an air brush. Both stencil and tape masking will help to achieve camouflage effects.

For The Ceramists

The air brush is a boon to the ceramist when it comes to applying a finish. To make the attractive "Man of La Mancha" ceramic shown in Fig. 8-11, proceed as follows:

Step 1. Clean greenware carefully and replace any necessary detail. Fire it in the kiln to the correct degree of temperature.

Step 2. After the piece cools, air brush the entire statue with white (Fig. 8-11A). When dry, carefully air brush the piece with silver.

Step 3. Air brush dark gray around the entire piece. This should be a controlled spray, allowing some silver to appear through the gray. This will add depth to the facial areas (Fig. 8-11B).

Step 4. Spray the facial area with a white/silver mixture to give detail. Then outline the other details with ivory and dark gray. To obtain a spray fine enough for this work, the color supply will be almost shut off and you should work very closely to the piece. Use short strokes instead of long ones.

Step 5. Highlight the eyes with white. Then spray the entire statue with a dull

A

B

C

D

Fig. 8-10: Camouflaging is an easy art thanks to an air gun.

141

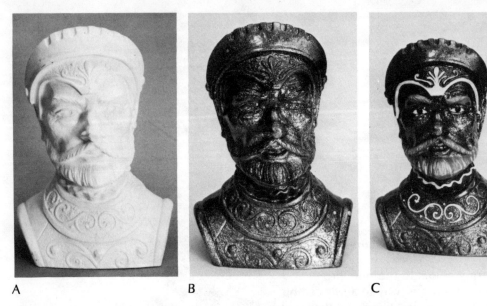

Fig. 8-11: Steps in applying a finish to a ceramic "Man of La Mancha."

matte spray to complete the job (Fig. 8-11C).

Cleaning and Care of An Air Brush

Proper care and common sense will produce a long service life. Keep your air brush and jar and/or color cup clean at all times and do not leave the air brush standing with materials for long periods, since this tends to gum up the internal feed channels. Keeping your air brush clean cannot be stressed too strongly; most of the problems encountered can be traced back to an air brush that has not been cleaned properly.

To clean the color chamber (Fig. 8-12), remove the jar and insert a bristle brush into the hole that the jar fits into, turning the brush to clear the chamber of the paint. Also, the chamber can be cleaned by inserting a cotton-tipped swab into the hole and removing the residual paint from the chamber. The latter operation is performed with the needle removed from the air brush.

After using a color cup with an air brush, use a bristle brush to swirl clean water or solvent inside the surfaces (Fig. 8-13A). Repeat until all the paint is cleaned away.

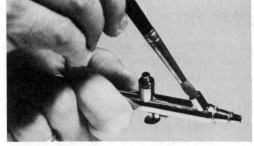

Fig. 8-12: Cleaning the color chamber.

Another method of cleaning the air brush (Fig. 8-13B) is to back flush it by filling the jar with clear water, or thinner, placing the air brush underneath the tabletop to prevent color splattering your work, and pulling back on the lever while pushing down for full air passage with your finger or soft cloth covering the tip. This will flush the color backward through the air brush, clearing and purging any leftover paint from the chamber and jar.

The needle should be cleaned with the proper solvent after each use (Fig. 8-14). When the needle is removed from the air brush for cleaning, make sure it is replaced properly and is snug against the tip. Do not

Fig. 8-13: Your air brush must always be kept clean.

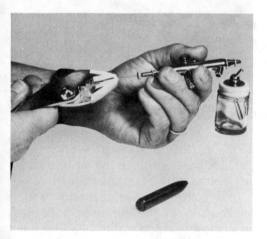

Fig. 8-14: If the needle is stuck in the air-brush, carefully loosen the needle chuck; then grasp the end of the needle with the pliers and twist in a counterclockwise direction to release the needle. Inspect for hardened paint, which causes the needle to bind. Hardened paint may be removed with thinner or extra fine sandpaper.

jam it into the tip. You will find that a residual stain will remain on the needle. One method of polishing it is to hold the needle flat on a worktable; then run a pink eraser the length of the needle (being extremely careful not to bend the tip), turning the needle slowly by rolling it toward yourself. This will remove all stains and paint particles from the needle body. Afterwards, be sure to remove all eraser particles by running the needle between your thumb and forefinger. When replacing the needle in the air brush, be sure to tighten the needle chuck firmly; the needle will not move and you will be unable to shut off the color flow. Always protect the tip of the needle; it may project beyond the spray regulator and be susceptible to bending.

Troubleshooting An Air Brush

Even with the best of care, an air brush will occasionally give you problems. The following are the most common troubles and how they can be solved.

1. **Grainy spray.** Caused by paint being too thick; add water sparingly to paint mixture, check the needle and regulator tip for dried paint, and check the air supply.

2. **Buckling paper.** Paint may be too thin; add pigment to thicken the mixture. Do not air brush as heavily in one area; move more rapidly or lessen your spray.

3. **Paint blobs at ends of stroke.** Paint is being sprayed before moving the hand and the movement is stopped before shutting off the paint flow.

4. **Flared ends.** Caused by turning the wrist while air brushing; the whole forearm should move horizontally across the paper.

5. **Centipedes.** Produced when spraying too much paint too close to paper. If a fine line is desired, lightly pull back on the front lever.

6. **Splattering.** Brought about if the needle is allowed to snap back into the tip. Always release the lever gently. Check for dried paint on needle or tip.

7. **Curved stroke.** Caused by arcing arm too close to the paper; arm should always be parallel to work, unless this effect is desired.

8. **Restricted spray.** Can be created by

spray regulator being screwed too tightly into head; open up a turn or two.

9. **Bubbles through color cup.** The spray regulator might be turned out too far; turn it back in a few turns. Color cup stem may be clogged.

10. **Color spray cannot be shut off.** Tip may be clogged; this is recognized by a "spongy" feel when needle is set into tip. A reamer can be used to clean out dry, gummy color by pushing it gently into the tip, slowly removing it in a rotating fashion, then pushing it gently back into the tip. Repeat until residue is out of the tip. Run clear water through the air brush. Take extreme care throughout this operation.

11. **Spitting.** Caused by residue on the needle or in the color cup. Paint may be too thick to operate properly. *Teflon*® head seal may not be seated properly.

Other Crafts

One of the most interesting of craft tools that works in conjunction with your air compressor is the sandblast gun, described in Chapters 2 and 5. With the proper abrasive, it can be used to give a driftwood effect to wood pieces, clean and polish silver and brass objects, and etch glass.

When driftwooding wood pieces, a sand grit size of 30 (16-45 mesh) should be used. The abrasive action of the sand will "eat" away the soft portions. It is possible to "age" furniture pieces (Fig. 8-15) and give wood finishes an antique effect. A sand grit of about 70 to 90 (40-140 mesh) should be used for most antiquing. Increasing the pressure to the sandblasting gun will speed up the eating or aging process. But never subject the canister to a pressure in excess of 50 psi when using the pressure feed mode.

When etching glass, a fine grit of 70 to 90 mesh should be used for the rough work, while glass beads should be employed for the final finishing operation. The portion to be etched requires careful masking (Fig. 8-16). Soft metals, such as aluminum and copper, can be given an "etched" effect by using a coarse sand or aluminum oxide with a grit size of 50 to 80. Working with the sandblast gun, like the air brush, you are limited only by your imagination.

To localize abrasive scattering and if reclamation of the abrasives is desired, an enclosure, such as the one described and illustrated on page 100, may be helpful.

Fig. 8-15: "Aging" a footstool by sandblasting.

Fig. 8-16: Masking glass in preparation to sandblasting.

FOR HOME WORKSHOP ENTHUSIASTS

For the do-it-yourselfer who has a complete workshop, a spray shop booth is a necessity. Some of the advantages of a spray booth are:

1. Segregates spraying operations in an enclosure for cleanliness and safety.
2. Reduces fire and health hazards.

3. Promotes cleaner conditions both for the spray operator and for the object being sprayed. Provides an area that can be kept clean more easily and also more effective dust control.

4. In a spray booth equipped with adequate lighting, better control of finishes is possible. It is the best way a perfectionist in finishing can achieve the results he/she wants for shop projects.

The ideal home shop installation is the separate finishing room, equipped with a turntable, paint cabinet, and an exhaust fan. Second choice is a spray booth located near a basement window to accommodate an exhaust fan. The spray booth need not be large—a 5 by 5 foot area will handle most projects. A booth in the corner is made with either doors that swing out to form the sides (Fig. 8-17), or a shower curtain hung on a rod from the ceiling. For safety in a paint spraying booth, if you choose the latter, the curtain should be fire-proofed. This is how: to 100 ounces of water add and dissolve 6 ounces of borax and 5 ounces of boric acid. Thoroughly soak the curtain in the warm solution several times in order to get maximum protection. The flameproofing solution is water-soluble; therefore after each washing it must be impregnated again for its further protection.

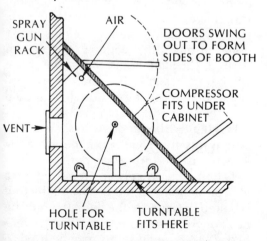

Fig. 8-17: A typical corner spray booth suitable for the average home workshop.

Turntables of simple construction are used in spray booths to facilitate rotation of the work while spraying. A suitable turntable eliminates a lot of body movement as well as handling of the work during the spraying process. The turntable and worktable shown in Fig. 8-18 can be placed on sawhorses when in use; these tables can be hung on the workshop walls when not being used. If desired, the height of the

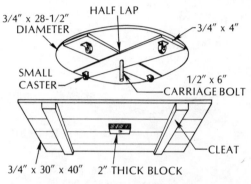

Fig. 8-18: A turntable is an aid to good work.

worktable can be made variable by suspending it with a rope or clothesline from hooks or other convenient overhead support. Another way to vary the height is to use a pedestal turntable, such as the one shown in Fig. 8-19. The 3/8 inch diameter pipe on the pedestal slides into the 3/4 inch pipe of the base and is held at the desired height by the collar-thumb screw arrangement. Two styles of bases are also illustrated. The casters permit the turntable to be easily moved about the booth, while the all-pipe base is more solid.

Small wooden articles can be readily sprayed if speared on an icepick, a brad point, or other special holding device, such as illustrated in Fig. 8-20. Objects without bases can sometimes be suspended from the ceiling.

An adjustable backstop stand, such as shown in Fig. 8-21, will prove very satisfactory. This stand uses newspapers or paper rolls and provides a suitable surface for testing the spray gun as well as a backstop when spraying the work. Remember that

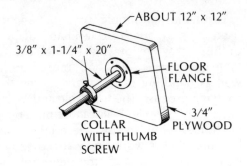

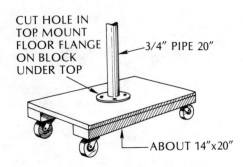

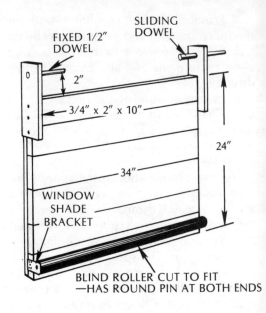

Fig. 8-21: A typical home workshop spray booth backstop.

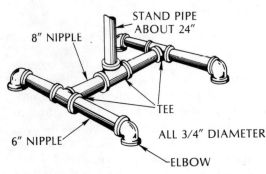

Fig. 8-19: A pedestal turntable for spraying small items.

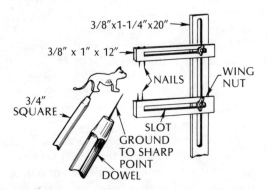

Fig. 8-20: Holding devices for spraying small items.

the home-built floor-type booth may accumulate a certain amount of waste material after the spraying has been completed. This should be removed immediately, since cleanliness is important to good spray results. When spraying is confined to an area, it is advisable to have a portable fan secured in a frame which can be placed in a window to provide proper ventilation. Spraying without a fan is best done with two basement windows open to provide ventilation. It is convenient to work directly under a window since the gun cleaning can be done through the window opening, if desired. The light should come from above whenever possible, but there are times when some form of spot floodlight should be at hand for side lighting.

Frequently, the serious workshop enthusiast wants the compressor permanently located. To perform such permanent installation, the wheels are removed or a vertical standing unit is used, and the unit is bolted or lagged to the floor. It should be positioned on the floor so that it rests evenly (Fig. 8-22). Solid shims may be used to level the unit. In bolting or lagging down,

Fig. 8-22: Leveling a compressor before bolting or lagging it permanently in place.

do not tighten so that strain is imposed.

When locating a compressor permanently, it should be placed as close as possible to the point or points where the compressed air is to be used. Make sure the spot is dry, clean, cool, and well ventilated. Keep it away from areas which have dirt, vapor, and volatile fumes in the atmosphere which may clog or gum the intake filter and valves, causing inefficient operation. The flywheel side (the belt guard) is usually placed toward the wall, but in no case should it be closer than 12 inches from the wall. Allow space on all sides for air circulation and for ease of of normal maintenance. The motor and switches should be wired by a competent

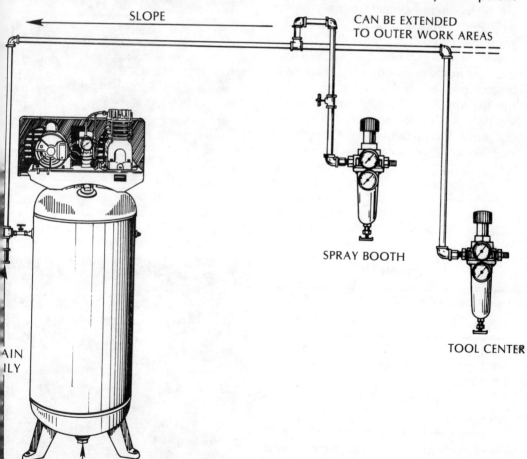

Fig. 8-23: A permanent compressor/air line arrangement such as this is the ultimate for any woodworker.

electrician in accordance with the National Electrical Code. Make certain that the motor has the same current characteristics as the power line to which it is connected. Correct type and size of wiring and full power supply will assure sufficient current during starting and other peak load periods.

Where the main air line is permanent or stationary for a distance of 10 feet or more, the line is usually made of pipe, replacing the conventional air hose. Copper or plastic pipe is the first choice in the homeshop because they are the easiest for the do-it-yourselfer to install, though galvanized pipe is probably used most often in professional establishments. The main line should slope slightly toward the tank or toward the filters in the line. Stationary piping facilitates connection of extractors and T joints for take-off stations. The take-off station or tool station, is where a T joint is placed in the main line to gain access to the compressed air (Fig. 8-23). Remember that the compressed air distribution system should be of sufficient pipe size to keep the pressure drop between the supply and the take-off station to a minimum. Also locate the filter or air transformer at the take-off station. Actually, there should be *at least* two take-off stations: one for the spray and the other at the workbench. The following table gives the recommended sizes:

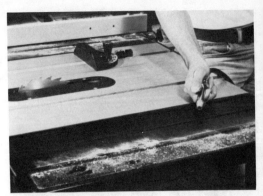

Fig. 8-24: Cleaning a table saw with a blow gun.

Fig. 8-25: A spot pinner/bradder at work. The pins and brads are driven well into the work.

Main Air Lines

hp	Lengths	Pipe dia.
1/3 & 1/2	All	1/4"
3/4 & 1	All	1/2"
1-1/2 & 2	All	3/4"
3 & 5	Up to 200'	3/4"
3 & 5	Over 200'	1"

A source of compressed air is almost indispensable in the home. When cleaning equipment such as a table saw (Fig. 8-24), drill press, shapers, or other tools which

Fig. 8-26: If your hobby is upholstering, an air driven stapler is a must.

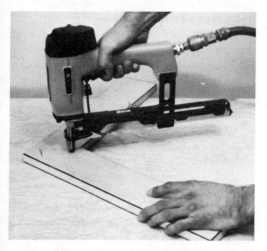

Fig. 8-27: An air-driven corrugated tool is excellent for fastening miter (left) or butt (right) joints.

cannot be immersed in a liquid or cleaned otherwise, an air compressor will finish the job in seconds.

For the crafts and woodworker, there are several air-powered tools to be considered. For example, there is the spot pinner/bradder (Fig. 8-25) which is ideal for fastening decorative trim, holding picture frames together, and for upholstery work. The stapler illustrated and described in Chapter 5, as well as smaller units (Fig. 8-26), can also be used in upholstering. The corrugated fasteners are an easy way to create strong bonds between wood pieces and they can be easily driven with an air tool such as shown in Fig. 8-27. In addition to

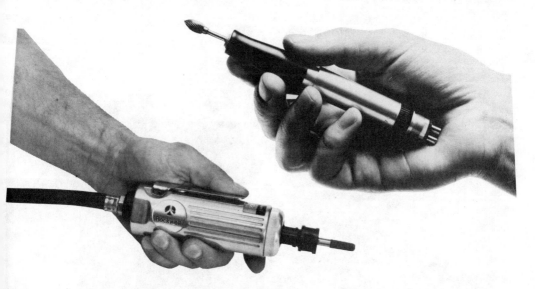

Fig. 8-28: (A) The die grinder is extremely valuable for woodcarving and sculpture. It weighs only a pound, yet the burr twists at a free-speed of 22,000 rpm; (B) the dental laboratory handpiece, at 25,000 rpm, does the same job as the more common flexible-shaft grinder, but it is less tiring to control.

the standard shop tools mentioned earlier in the book, there are a large variety of special-purpose routers and laminate trimmers, which could be most useful in general woodworking. There is also a huge variety of pneumatic clamping and positioning machinery for gluing sub-assemblies and for holding parts in position. And for the woodworking specialist, such as a sculptor or carver who works with green logs and large stacked forms, there are air-powered adzes, cutting burrs, die grinders (Fig. 8-28), and carving gouges that will save hours of handcutting, gouge-tapping, and sanding. True, while these specialty tools are designed and manufactured for industry, there is nothing to stop the serious woodworker from having them in his/her shop.

FOR OUTDOORSPEOPLE

For the boat lover, the air compressor and its spray gun accessory can paint the hull whether wood, metal, or plastic. Start at the gunwale and spray in sections down to the keel. Two or three light coats of marine paint will give better protection than one heavy coat. Your spray gun will shorten hours of brushing on a deck paint or a spar varnish. To lessen preparation time, an air-powered sander can be used.

To remove rust from a boat trailer or other marine equipment, use your sandblasting equipment. A rust resistant primer should be applied over the bare metal, followed by at least two coats of marine paint.

If you own your own landing dock, the spray gun can be used not only to paint it but also to rid it of fungus by spraying various anti-bacterial preparations. Boat sheds or other water-based structures are susceptible to infestations of many kinds of insects, and often become mildewed. An occasional once-over spraying with the proper insecticides will eliminate such problems. By reducing the time spent on the unpleasant task of getting your boat ready for the water, you will be able to do what you really want—boating, sailing, and fishing. By the way, if your boating is done

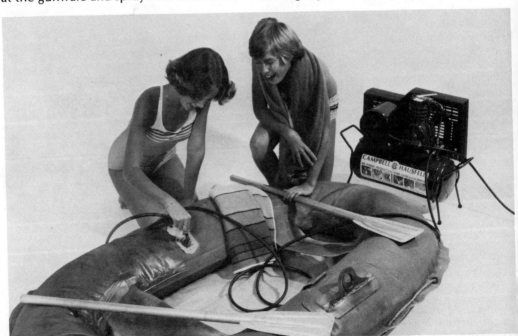

Fig. 8-29: It is no problem to get inflatable craft in the water; in fact, it is even child's play.

in one of those new inflatable crafts, it is no problem to get it in the water quickly (Fig. 8-29).

If campers and recreational vehicles are your hobby, you can avoid the delay and expense of having yours painted. Wash it down with water using your sandblast equipment, then repair any damaged or worn areas as described in Chapter 7, and spray paint it as you would any other structure. The shock absorbers and the plumbing system can be charged with the compressor, and when it comes time to winterize your vehicle, use the blow gun to blow out the water system to prevent freezing (Fig. 8-30). And while traveling, be sure to have a portable air tank available to inflate tires, air mattresses, or a boat.

All in all, an air compressor plus its many tools is one of the most versatile pieces of equipment that any home owner, do-it-yourselfer, or hobbyist can own.

Fig. 8-30: A blow gun, used to force the water out of the pipes, is an important part of winterizing the trailer.

INDEX

Abrasive materials, 20, 99-100, 144
Accessories, 10-12, 27, 54-60
   troubleshooting, 58-60
Aging wood, 144
Air brush paints, 138-139
   brushes, 18-19, 130, 131, 137-144
   brushing, 18, 27, 130, 137-144
      cleaning, 142-143
      troubleshooting, 143-144
   caps, 14-15, 64-65
   driven bradders, 108, 111, 149
      caulk guns, 22, 26, 81, 83-86, 110-111, 112-113
      chisels, 26, 127-128
      corrugated fasteners, 149-150
      die grinders, 26, 128, 150
      drills, 1, 26, 105, 125
      grinders, 26, 126, 127
      impact wrenches, 24, 25, 26, 105, 121-124
      nailers, 26, 108, 111
      ratchet wrenches, 26, 124
      sandblast guns, 20-22, 26, 81, 89, 100-102, 128, 144
      sanders, 26, 89, 98, 125-127
      staplers, 26, 106-107, 149
      wrenches, 121-124
   filters, 10-11, 12, 37, 57, 59
      troubleshooting, 59
   hoses, 9, 12
   motor care, 129
   pressure, 1, 6
   regulators, 4, 10, 12, 54-56, 121-123
      adjusting, 56
      assembling, 55-56
      servicing, 55-56
      troubleshooting, 58-59
   spraying ceramics, 141-142
      models, 139-141
   tanks, 1, 3-4, 7-9, 25, 41, 58, 151
   tools, 1, 9, 10, 13, 23-25, 98-99, 105, 121-129, 149-150
   transformers, 11, 12, 57-58, 59-60
      troubleshooting, 59-60
American Society of Mechanical Engineers, 8
Angle air caps, 15

Belt alignment, 37-38
   filters, 11, 56, 59
      troubleshooting, 59
   guard, 12
Bleeder air brushes, 18
   sandblast guns, 20, 101-102
   spray guns, 13
Blow guns, 23, 26, 27, 105, 109, 110, 128, 148-149, 151
Bradders, 108, 111, 149
Break-in period, 36, 52
Brush/weed killers, spraying of, 102, 104-105

Camouflaging techniques, 141
Caulk guns, 22, 26, 81, 83-86, 110-111, 112-113
   loading, 83-84
   using, 83-86
Caulking, 1, 22, 81-86, 110-111
   applying, 83-86
   bathtubs and showers, 110-111
   compounds, 81-82
   locating, 82-83
Check valves, 9, 39-41
Chisels, 26, 127-128
Cleaning air brushes, 142-143
   blast, 99-102
   equipment, 22-23
   material tanks, 74-75
   spray guns, 73-75
Coded tanks, 8-9
Compact compressors, 1, 4, 19, 21, 27, 46, 51
Compressor, permanent location of, 146-148
   operation, 29-32
   ratings, 4-6
   safety, 12
   stages, 6-7
   types, 1-4
Compressors, diaphragm, 2, 4, 27, 51-54
   oilless, 4, 27, 46-51
   piston, 2-4, 29-46
   troubleshooting, 41-46, 50-51, 53-54
Contour putty, 139-140
Convertible air brushes, 18
   sandblast guns, 101-102
   spray guns, 14, 16, 18
Corrugated fasteners, 149-150
Couplers, 9
Crabgrass killers, spraying of, 102
Craft series, 27
Crankcase plugs, 32
Cubic feet per minute (cfm), 5-6
Customized car painting, 136

Diaphragm compressors, 2, 4, 27, 51-54
   lubrication and break-in, 52
   maintenance, 52-53
   operational tips, 51-53
   power requirements, 51
   troubleshooting, 53-54
Die grinders, 26, 128, 150
Displacement cfm, 5
Draining air tanks, 10, 36-37
Drills, 1, 26, 105, 125
   use of, 125
Drive belt alignment, 37-38
   belts, 37-38

Electrical power requirements, 26, 32-35, 47, 51, 147-148
    problems, 32
Etching glass, 144
Extension cord safety, 34-35
    cords, 12, 34-35
Exterior painting, 86-98
    surface preparation for, 89-92
    when to do, 92
Exterior paints, application of, 92-98
    estimating quantity of, 88-99
    selection of, 86-88
    spraying of, 94-96
External-mix nozzles, 14, 15, 16

Fan pattern caps, 15
Farm and field series, 28
Finishing sanders, 126-127
Fishing lures, painting of, 137-138
Floor tiles, applying, 113
Flywheels, 37
Free air cfm, 5-6
Fungicides, spraying of, 102, 104
Furniture finishing, 97-98, 119-120

Garden spraying, 102-104
Gasoline engines, 35-36
Glazing putty, 130, 134, 139-140
Grease guns, 26, 105, 128
Grinders, 26, 126, 127
Grounding requirements, 33, 34

Home series, 27
Hoppers, sand, 20, 21-22
Horsepower, 4-5, 6

Impact wrenches, 24, 25, 26, 105, 121-124
    use of, 123-124
Inflating items, 1, 24-25, 105, 151
Inflator kit, 24, 25, 26, 27, 81, 137
Insect spraying, 103-105
Insecticides, 20, 102, 103-104, 109, 150
Intake air filters, 37
Interior painting, 113-119
    surface preparation for, 115-116
Interior paints, application of, 117-118
    estimating quantity of, 115
    selection of, 113-114
Internal-mix nozzles, 14, 15, 18

Lubricated piston compressors, see *Piston compressors*
Lubrication of compressors, 32, 129
Lubricators, 10, 11-12, 129

Maintenance, diaphragm compressors, 52-53
    oilless compressors, 47-49
    schedule, 36
    series, 27

Masonry surfaces, cleaning, 1, 87-88
    painting, 91-92
Material tank cleaning, 74-75
    hookup, 62
    spraying from, 65-66
    tanks, 16, 17-18, 26, 61, 62, 65-66, 93
Multi-stage compressors, 6

Nailers, 26, 108, 111
National Electrical Code, 34, 148
Needle sealers, 26, 98-99
Non-bleeder air brushes, 18
    sandblast guns, 21, 101-102
    spray guns, 13

Occupational Safety and Health Act, 8-9
Off/automatic switch, 4, 33
Oil requirements, 32, 36
Oilless piston compressors, 4, 27, 46-51
    maintenance of, 47-50
    motor for, 50
    operation of, 47
    power requirements for, 47
    troubleshooting of, 50-51

Pad sanders, see *Finishing sanders*
Paint material stirring, 92-93, 94
    thinning, 72
Painting appliances, 119
    bamboo, 98
    basement walls, 119
    bicycles, 137
    cabinets, 118, 120
    cars, 1, 133-136
        customized, 136
    ceiling and walls, 117-118
    ceramics, 141-142
    doors, 118
    exterior, 86-98
    fences, 97
    finishing lures, 137-138
    floors, 96
    furnaces, 119
    furniture, 119-120
    interior, 113-119
    masonry surfaces, 91-92
    metal, 92, 102, 133-136
    models, 139-141
    outdoor furniture, 97-98
    radiators, 119
    rattan, 98
    shingles and shakes, 96
    siding, 94-95
    shutters, 96
    spray, 1, 10, 61-80, 113-119, 136
    storm sashes and screens, 96-97
    surface preparation for, 89-92, 115-116
    tools, 98
    trellis, 98
    trim and baseboards, 118
    T-shirt, 138
    wicker, 1, 98

Permanent location for compressor, 146-148
Pipe air lines, 9, 148
Piston compressors, 2-4, 29-46
    troubleshooting of, 41-46
Plastic filler, 130, 131
Plug adapters, 33-34
Plumbing tips, 110-111
Pneumatic tools, see *Air tools*
Portable air tanks, 25, 58, 151
Pressure, 6, 19
    feed guns, 15, 20-21, 100, 103
    regulators, 4, 10, 12, 54-55, 58-59
        adjusting, 56
        assembling, 55-56
        servicing, 55-56
        troubleshooting, 58-59
    switch, 3, 4, 33, 38-39
    tanks, 16, 17-18
    washers, 22-23, 89-90, 96, 102, 128

Quick-connect couplers, 9

Ratchet wrenches, 26, 124
Receivers, see *Tanks*
Refinishing cars, 130-136
    furniture, 1, 119-120
Rental service, 108
Repairing minor dents, 130-131
    rustouts, 132-133
    scratches, 130
Respirators, 12, 61, 100, 109
Rotary scalers, 26, 98-99
Round pattern cap, 15
Rust, blast cleaning of, 99-102
    hand cleaning of, 98
    power tool cleaning of, 98-99
    removal, 1, 98-102, 150

Safety relief valves, 3, 28
Sandblast guns, 20-22, 26, 81, 89, 100-102, 128, 144
    bleeder, 20, 101-102
    non-bleeder, 20, 101-102
    pressure feed, 20-21, 100-102
    siphon feed, 20, 100
    troubleshooting, 102
    cleaning of rust, 99-102
Sandblasting, 19, 62, 99-102, 109
    enclosure, 100-101, 144
    equipment, 19-22, 100, 102, 150, 151
Sand hoppers, 20, 21-22
Sanders, 26, 89, 98, 125-126
    safety with, 127
    use of, 125-127
Scaffolding, 93
Selecting air brushes, 18-19
    compressors, 6, 18, 25-28
    exterior finishes, 86-88
    interior finishes, 113-114
    spray guns, 18

Single action air brushes, 19
    stage compressors, 6
Siphon canister hookup, 61-62
    feed guns, 15, 20, 61-62, 100, 101-102, 103
Spot putty, 132
Spray booths, 144-146
    guns, 10, 13-18, 26, 61-80, 102, 131, 137, 150
        cleaning, 73-75
        problem with, 78-80
        types, 13-14
        use of, 63-73
    nozzles, 14-15
    painting, 1, 10, 61-80, 113-119, 136
        masking, 71-73
        materials, 86-88, 113-114
        problems, 75-80
        safety, 61
        shielding, 71, 72-73
        stroke, 66-71
        surface preparation, 62, 89-92, 115-116
        when to, 92
    patterns, 14-15, 64-65, 72
    systems, 61-62
Spraying gardens and trees, 102-104
Staplers, 26, 106-107, 149
Stationary compressors, 9, 146-148

Tank compressors, 1, 3-4, 21, 27, 33, 46
    size, 7-9
Tanks, 1, 3-4, 7-9, 41
Terminal overload, 33, 51-52
Tire chucks, 25, 26, 27, 105, 128-129
Troubleshooting accessories, 58-60
    air brushes, 143-144
    diaphragm compressors, 53-54
    oilless compressors, 50-51
    piston compressors, 41-46
    spray gun problems, 75-80
T-shirt painting, 138
Two-stage compressors, 6-7

Unclogging pipes, 110
Unloaders, 39-41

Vinyl floor tile, 113
Viscosimeter, 63
Volumetric efficiency, 6

Wall coverings, applying, 112-113
    paneling, 112
Wallpaper removal, 115
Wire brushes, 99
Wiring requirements, 32, 33-34, 148
Wood surfaces, painting of, 86-87
    tile, 113
Wrenches, 121-124